HOW TO PASS

In full COLOUR

INTERMEDIATE 2
MATHS

Brian J Logan

HODDER
GIBSON
AN HACHETTE UK COMPANY

Orders: please contact Bookpoint Ltd, 130 Milton Park, Abingdon, Oxon OX14 4SB. Telephone: (44) 01235 827720. Fax: (44) 01235 400454. Lines are open from 9.00–5.00, Monday to Saturday, with a 24-hour message answering service. Visit our website at www.hoddereducation.co.uk. Hodder Gibson can be contacted direct on: Tel: 0141 848 1609; Fax: 0141 889 6315; email: hoddergibson@hodder.co.uk

© Brian J Logan 2006, 2009
First published in 2006 by
Hodder Gibson, an imprint of Hodder Education,
an Hachette UK Company
2a Christie Street
Paisley PA1 1NB

This colour edition published 2009

Impression number	5	4	3
Year	2012	2011	

Cover photo Dominique Sarraute/The Image Bank/Getty Images
Illustrations by Tech-Set Ltd, Gateshead
Typeset in 9.5/12.5 Frutiger by Tech-Set Ltd, Gateshead
Printed in Dubai

A catalogue record for this title is available from the British Library.

ISBN 13: 978 0340 974 117

CONTENTS

CONTENTS

INTRODUCTION TO INTERMEDIATE 2 MATHS

Welcome to How to Pass Intermediate 2 Maths!

The fact that you are reading this book shows that you want to pass the exam and hopefully the book will benefit you, whether you need help simply to pass the exam, or whether you are aiming for a top grade.

This book is designed to show you the type of questions which are most likely to appear in the exam and how to answer them. Details of the marks awarded are given along with hints on how to tackle questions and notes on how to avoid common errors.

As you study each chapter, please try the examples yourself **before** reading the solutions, and take seriously any advice that I have given regarding practising further examples. This book alone is not enough to enable you to pass the exam – you must prepare properly by further study and practice.

The Intermediate 2 Exam

The exam consists of two papers. Paper 1 is a non-calculator paper. It lasts for 45 minutes and is worth 30 marks. Paper 2 allows the use of a calculator. It lasts for 1 hour and 30 minutes and is worth 50 marks. So the total mark for the exam is 80.

Questions will be asked on three Units, each of which is divided into outcomes.

All students do Unit 1 and Unit 2. The third unit is **either** Unit 3 **or** Unit 4, the Applications of Mathematics, whichever option you have chosen.

Out of the 80 marks available in the exam, there are usually 27 marks on Unit 1, 27 marks on Unit 2 and 26 marks on the optional third unit.

In this book, if you are studying Unit 3, do not read Chapters 15 to 18. They do not apply to you. Similarly, if you are studying the Applications of Mathematics Unit, do not read Chapters 12 to 14, as they do not apply to you. Be careful when you are doing the Practice Papers in Appendices 1 and 2 that you do the questions from the Units you are studying.

Why is there an option for the third Unit? Well, Unit 3 would usually be studied by candidates who are hoping to continue with maths and possibly sit Higher Maths later on. The work in Unit 3 helps you to prepare for Higher Maths.

The Applications of Mathematics Unit is normally taken by students who are not going to continue with maths after they complete Intermediate 2.

Some schools or colleges may offer only one option for the third unit depending on their circumstances.

As I said, the exam is worth 80 marks. You will probably need to score 50% (40 out of 80) for a C pass, 60% (48 out of 80) for a B pass, and 70% (56 out of 80) for an A pass.

Around 65% to 70% of the marks in the exam are described as 'routine'. This means that they are at level C. You should do questions similar to these during the session as you work through the course.

The remainder of the marks, about 30% to 35% are non-routine, i.e. more difficult questions at A/B level. You should also regularly be exposed to these more difficult questions throughout the session.

During the session you will have to sit and pass Unit Tests at the end of each Unit.

If you are attending regularly and working properly, these tests should not prove to be a problem, and there is even a chance to re-sit them if necessary. However, beware!

The Unit Tests are testing basic competence. This means that they are testing whether you can do the simplest and most straightforward things in each Unit. Some students mistakenly think that the Unit Tests reflect the degree of difficulty of the final exam. This is not the case. The final exam is much more difficult, so don't get carried away by a high mark in your Unit Test. On the contrary, if you are not scoring high marks in the Unit Tests, it is a cause for worry and time to start working much harder.

Your Calculator

In Paper 2, you are allowed to use a calculator. It is essential that you have your own calculator, one with which you are very familiar. You will also need it as you work through this book. Because there are many different makes and types of calculator, it is important that several points are made now.

You can use either a *scientific* or a *graphical* calculator. A scientific calculator will be perfectly adequate and is probably simpler to use, although you must stick to what you use normally.

Because of different makes of calculator, I will mention several useful functions which you should be aware of and able to use.

Make sure the calculator is in the *degree* mode as this is necessary for questions on trigonometry. This is usually indicated by DEG or D on the screen. As a test, if tan 45° = 1, then you are in degree mode. If you are not in degree mode, make sure you know how to return to it.

You must be able to use the sin, cos and tan functions, including the inverse functions, usually $\sin^{-1}$, $\cos^{-1}$, and $\tan^{-1}$. We shall investigate these in detail in Chapter 8.

I recommend that you use the π key instead of 3.14 in circle calculations and will return to this later.

There is a key for squaring numbers, often x^2, which is useful. Try it to find $8^2 = 64$ as an example.

Another very useful key, as you will see when we study percentages, is the power/index key. This is often y^x or ^, and can be used to calculate powers quickly, such as 3^5. Try this using the power/index key. The answer is 243.

It is your job to get to know your calculator and to be able to use it accurately and consistently in class, at home and in exams. There are many keys on your calculator which you will never use in Intermediate 2. If in any doubt about your calculator, ask your teacher.

Studying for the Exam

As I mentioned, you would not be reading this book unless you wanted to pass Intermediate 2 Maths. All students say that they want to pass exams, but some would like to be able to pass them without having to study. Sadly this is not possible. So it is essential, not only that you study, but that you study sensibly.

As the Intermediate 2 course consists of Units divided into *outcomes*, it is easy to break down the course into manageable chunks to study. There are 12 outcomes if you are studying Unit 3, and 13 if you are studying the Applications of Mathematics Unit.

Some outcomes are shorter than others – the outcome on percentages in Unit 1 is fairly short and can be revised and practised fairly quickly, whereas the trigonometry outcome in Unit 2 will take longer.

You should make a list of all the outcomes and plan your studying to fit them all in. It is important, too, that you re-visit outcomes you have studied earlier in order to refresh your memory. Each outcome is dealt with in a separate Chapter in this book.

Finally, here is some advice about doing Past Papers. I consider it extremely important that you do as many Past Papers as possible. The first Intermediate 2 exam took place in the year 2000 and there has been one every year since with the exception of 2002 when there were two exams (an extra one took place in January).

You probably won't have time to do them all in class by the time your teacher has worked through the syllabus, completed Unit Tests, and you have sat preliminary exams. So it is up to you to get hold of Past Papers, practise them at home and have them checked. You can do certain questions during the session as you complete an outcome. Ask your teacher for advice on which questions to try.

I can also recommend 'Intermediate 2 Maths Practice Papers' by Peter Westwood, published by Hodder Gibson, for further practice.

As you work through Past Papers you will find that certain types of questions re-appear most years and as you practise them your confidence will increase.

In the Appendix at the end of this book you will find, in addition to some hints on sitting the exam and some useful vocabulary, a chart listing the frequency with which all the topics in the course have appeared in the exam since 2000. This should be useful in helping to focus your studying.

A fan once said to the great American golfer Arnold Palmer that he was a *lucky* golfer. He replied 'It's a funny thing, the more I practise the luckier I get'.

It's the same with studying maths. I wish you good luck with your studying and preparation, and also with the exam itself.

REVISION TOPICS

 It will be most useful to you to read through this Chapter to remind yourself of the topics you are expected to know before you start Intermediate 2 Maths. All of these topics will occur in examples later in the book.

Rounding

Remember

There are three ways of rounding:

◆ to the nearest unit
◆ to a number of decimal places
◆ to a number of significant figures.

Example

a) Round 57 837 to the nearest thousand.
b) Round 6·371 28 to one decimal place.
c) Round 53 217 to two significant figures.

(Solution) a) 58 000.
  b) 6·4   (6·371 28 has five decimal places, i.e. five numbers after the decimal point. To round it to one decimal place you don't need 7128, i.e. 6·3~~7128~~. As the number after the 3 is 5 or more, add 1 to the 3 giving 6·4 which has one decimal place).
  c) 53 000   (53 217 has five significant figures. To round it to two significant figures you don't need 217, i.e. 53~~217~~. As the number after the 53 is less than 5, leave the 53 and add three zeros. The zeros are not significant, they show that the 53 represents 53 thousand).

A number such as 0·007 635 has four significant figures. The zeros at the start are not significant. This number would be 0·007 64 rounded to three significant figures.

Percentages

Remember

Can you calculate 38% of £240, using a calculator? Remember that $38\% = \dfrac{38}{100} = 0{\cdot}38$, therefore 38% of $£240 = 0{\cdot}38 \times 240 = £91{\cdot}20$.

Remember, too, that $33\frac{1}{3}\% = \frac{1}{3}$ and $66\frac{2}{3}\% = \frac{2}{3}$.

Example

A caravan decreased in value from £15 000 to £14 100 in one year.
Find the percentage decrease.

(Solution)   Actual Decrease $= £(15\,000 - 14\,100) = £900$.

$$\text{Percentage Decrease} = \frac{actual\ decrease}{original\ value} \times 100 = \frac{900}{15\,000} \times 100 = 6\%.$$

Area and Volume

Formulae to Learn

You must *memorise* the following formulae:

Area of a square $= l^2$;   Area of a rectangle $= lb$;   Area of a triangle $= \frac{1}{2}bh$;
Volume of a cube $= l^3$;   Volume of a cuboid $= lbh$;   Volume of a prism $= Ah$.

(You should already have met them often. Some of them will appear again in the next few Chapters.)

The Circle

Formulae to Learn

There are two formulae to memorise.

Area of a circle, $A = \pi r^2$ and Circumference of a circle, $C = \pi d$.

Remember that the circumference of a circle is the same as its perimeter, i.e. the distance around the outside.

Example A circle has diameter 9 centimetres. Calculate
a) its area   b) its circumference

(Solution) a) Area $A = \pi r^2 = \pi \times 4\cdot5^2 = 63\cdot6\,\text{cm}^2$. (remember to halve 9 for the radius)
b) Circumference $C = \pi d = \pi \times 9 = 28\cdot3\,\text{cm}$.

You can use either the π button or $3\cdot14$ in the calculations.

Pythagoras' Theorem

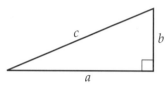

$$c^2 = a^2 + b^2$$

We can use Pythagoras' Theorem in a *right-angled triangle* to calculate the length of a side when we know the lengths of the other two sides.

If we are finding the **hypotenuse** (the longest side opposite the right angle), square and **add** the other two sides. If we are finding one of the two shorter sides, square and **subtract** the other two sides.

Example

Find the lengths of the sides x and y in these right-angled triangles.

a)

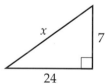

b)

(Solution) a) $x^2 = 7^2 + 24^2$
    $= 49 + 576$
    $= 625$
    $\Rightarrow x = \sqrt{625}$
    $= 25$.

b) $8\cdot2^2 = y^2 + 5\cdot1^2$
    $\Rightarrow y^2 = 8\cdot2^2 - 5\cdot1^2$
    $= 67\cdot24 - 26\cdot01$
    $= 41\cdot23$
    $\Rightarrow y = \sqrt{41\cdot23}$
    $= 6\cdot4$.

Trigonometry

You should be able to do basic trigonometry calculations in a right-angled triangle, finding either the length of a side or the size of an angle.

Use a calculator and make sure it is in **degree mode.**

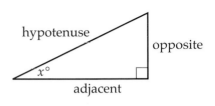

$$\sin x° = \frac{\text{opposite}}{\text{hypotenuse}}$$

$$\cos x° = \frac{\text{adjacent}}{\text{hypotenuse}}$$

$$\tan x° = \frac{\text{opposite}}{\text{adjacent}}$$

 You must remember these as the formulae are not given in the formulae sheet. By taking the first letters in the order above, you can use SOH CAH TOA as a memory aid.

When you have a trigonometry problem in a right-angled triangle, remember that the hypotenuse (H) is opposite the right angle, the opposite (O) is opposite the marked angle, and the adjacent (A) is the remaining side. You should then be able to decide whether to use sine, cosine or tangent.

Example

Find the lengths of the sides marked x in the diagrams below.

a)

b)

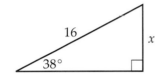

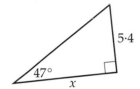

(Solution)

$$\sin 38° = \frac{x}{16}$$

$$\Rightarrow x = 16 \times \sin 38°$$

$$\Rightarrow x = 9.85.$$

$$\tan 47° = \frac{5.4}{x}$$

$$\Rightarrow x \times \tan 47° = 5.4$$

$$\Rightarrow x = \frac{5.4}{\tan 47°}$$

$$\Rightarrow x = 5.04.$$

Example

Find the size of the angle marked $a°$ in the diagram below.

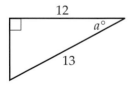

(Solution) $\cos a° = \dfrac{12}{13}$

$\qquad = 0.923$ (now use $\cos^{-1}$ or equivalent to find angle)

$\Rightarrow a \quad = \cos^{-1} 0.923 = 22.6.$

 Check all of the working carefully and make sure you follow each step.

Statistics

You should be able to construct and interpret *scattergraphs* and *stem and leaf* diagrams, as well as knowing about the different *averages* used in statistics. We shall revise all of these topics.

Scattergraphs

Example

A group of students was given short tests in English and French. Their results are shown in the table below.

Student	A	B	C	D	E	F	G	H	I	J	K	L
English mark	8	9	6	2	7	9	5	9	6	7	3	4
French mark	6	7	6	1	5	8	6	9	4	6	2	4

a) Illustrate this data on a scattergraph.
b) Draw a best-fitting line on the scattergraph.

(Solution)

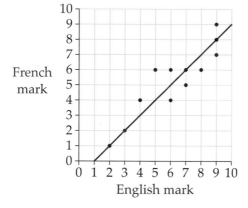

Make sure you put clear scales on both axes and label both axes. The first set of numbers in the table, English marks in this case, should form the horizontal axis.

When drawing the best-fitting line, it should follow the general path of the dots and be drawn so that there are the same number of dots above the line as there are below it.

Stem and Leaf Diagrams

Example

The ages of the inhabitants of Glebe Street are shown below.

```
14  59  38  25   7  51  42  36  40  36  23  19
 8  23  47  35  39   2  50  33  36  28  17  12
20  28  16  17  29  30  35  58  24  50  18  22
```

Illustrate this data on a stem and leaf diagram.

Example continued ➤

Example *continued*

(Solution)

0	2	7	8						
1	2	4	6	7	7	8	9		
2	0	2	3	3	4	5	8	8	9
3	0	3	5	5	6	6	6	8	9
4	0	2	7						
5	0	0	1	8	9				

$n = 36$ $3\,|\,5$ represents 35

Are you clear about illustrating data on a stem and leaf diagram? Remember to say what *n* is, and to provide a key to the diagram. Some students miss out part of the data through carelessness. Always check, by counting, that you have included all the 'leaves' in the diagram.

Averages

You should know about three important 'averages' in statistics – the *mean*, the *mode* and the *median*.

Key Points

For any data set, the mean $= \dfrac{Sum\ (total)\ of\ values}{number\ of\ values}$;

the mode = the most frequent value;

the median = the middle value in an ordered set.

Example

Find the mean, mode and median of

6 6 8 9 6 6 8 5 3 9 10 11 6 4 8.

(Solution)

$$\text{Mean} = \frac{total\ of\ values}{number\ of\ values}$$

$$= \frac{(6+6+8+9+6+6+8+5+3+9+10+11+6+4+8)}{15}$$

$$= \frac{105}{15}$$

$$= 7.$$

Mode = 6 (the most frequent value).

For the median, write numbers in order:

3 4 5 6 6 6 6 6 8 8 8 9 9 10 11

Median = 6. (the middle number is the 8th number along, underlined)

You should also know how to calculate the *range*, a measure of the *spread* of the numbers.

Range = Highest − Lowest

In the example above, the range = 11 − 3 = 8.

We have now covered some important aspects of revision. They will appear again in later Chapters. Now we shall start looking at the Intermediate 2 syllabus in detail.

PERCENTAGES

In this outcome, you are expected to carry out calculations involving percentages in context. This involves *appreciation* and *depreciation* (and includes compound interest). This topic will appear in the calculator paper, so have your calculator ready.

The usual exam questions involve increasing a quantity by a given percentage over a period of time (appreciation), or decreasing a quantity by a given percentage over a period of time (depreciation).

Such questions are usually worth three marks. A typical example is shown below, along with two methods of doing it.

Appreciation

Example

The population of Ferryport is 36 000. It is expected to increase by 8% per annum. What will the population be in 3 years' time?
(Give your answer to the nearest thousand.)

(Solution)   There are two ways of doing this. Remember that *per annum* means each year.

Method 1 – year by year

1st Year   Increase = 8% of 36 000 = $0 \cdot 08 \times 36\,000 = 2880$
New Population = $36\,000 + 2880 = 38\,880$.

2nd Year   Increase = 8% of 38 880 = $0 \cdot 08 \times 38\,880 = 3110$ (rounded)
New Population = $38\,880 + 3110 = 41\,990$.

3rd Year   Increase = 8% of 41 990 = $0 \cdot 08 \times 41\,990 = 3359$ (rounded)
New population = $41\,990 + 3359 = 45\,349$.

Answer: 45 000 (to the nearest thousand).

Method 2 – by repeated multiplication
(This method uses the fact that when you increase a quantity by 8%, the new value is 108% of the old value. $108\% = 1 \cdot 08$ as a decimal, so you can simply multiply the original quantity by $1 \cdot 08$ for each year, i.e. by $1 \cdot 08 \times 1 \cdot 08 \times 1 \cdot 08$ or $1 \cdot 08^3$.)

Hence new population = $36\,000 \times 1 \cdot 08^3 = 45\,349 \cdot 632 = 45\,000$ (to the nearest thousand).

To calculate $36\,000 \times 1 \cdot 08^3$, either use the power/index key on your calculator, usually y^x or ^, or key in $36\,000 \times 1 \cdot 08 \times 1 \cdot 08 \times 1 \cdot 08$.

Method 2 is more efficient as it requires only one calculation, reducing the chance of making a mistake. Also, Method 1 would be completely unsuitable if the number of years was much greater, say 8 years. However, you must use the method you are more comfortable with. Don't forget to round your answer to the nearest thousand. Many students forget and lose a mark. If you are using the first method, don't round to the nearest thousand until the end of your calculations.

Three marks would be awarded for this type of question. One for knowing how to increase by 8%, one for doing this for three years, and one for doing the calculations and rounding correctly.

Common Mistakes

Never calculate 8% of 36 000, multiply it by 3, and add it on!
This is the wrong method and you will get no marks.

Example

Calculate the compound interest on £12 500 invested for 4 years at 3·8% per annum.

(Solution)   Using Method 2, to increase by 3·8%, we add 3·8 to 100, giving 103·8% ($= 1·038$ as a decimal).

The total amount after 4 years $= 12\,500 \times 1·038^4 = £14\,511·07$ (to the nearest penny).
Hence the compound interest $= 14\,511·07 - 12\,500 = £2011·07$.

If you still prefer to use Method 1, try it yourself, check the answer, and then decide which method is quicker!

Depreciation

Now we look at depreciation in which a quantity is reduced by a given percentage over a period of time. (Again we shall use Method 2.) Suppose a quantity is to be reduced by 8%, then its new value will be $(100 - 8)\% = 92\%$ of its old value, so you can simply multiply the original value by 92%. Written as a decimal this is 0·92.

(As before you can still use Method 1 and check your answer.)

Example

The value of a caravan drops by 12% each year. The caravan cost £28 000 when new. What will its value be after 3 years?
Give your answer to the nearest thousand pounds.

(Solution)   Value after 3 years $= 28\,000 \times 0·88^3 = £19\,081·22$.

Answer: £19 000 (to the nearest thousand pounds).

The next example shows how Method 2 can be adapted to a question in which the percentage changes over a period of time.

Example

Sean is having difficulty with his central heating. At 7 a.m. the temperature in his house is 16 °C. Over the next three hours, the temperature rises by 6%, drops by 4%, then rises by 15%. Calculate what the temperature will be at 10 a.m.

(Solution)   New temperature $= 16 \times 1.06 \times 0.96 \times 1.15 = 18.7$ °C.

Look again at the working for this example carefully and make sure you understand it.

Next we shall consider a two part question. If you do not get an answer to part (a), however, you will be unable to start part (b), so it is very important you study this and learn how to get started.

Example

Mary's pay increased from £15 000 to £15 600 in one year.

a)   What was the percentage increase?

b)   If her pay continued to rise at this rate, what would her pay be after a **further** 3 years? Give your answer to the nearest hundred pounds.

(Solution)   a)   Actual increase $= 15\,600 - 15\,000 = £600$.

Percentage increase $= \dfrac{600}{15\,000} \times 100 = 4\%$.

b)   Pay after a further 3 years $= 15\,600 \times 1.04^3$
$= £17\,547.88$
$= £17\,500$ (to the nearest hundred pounds).

Hints *and* Tips

In this question part (a) would be worth one mark, and part (b) three marks.

The solution to part (a) should remind you that to find a percentage increase (or decrease) you divide the actual increase (or decrease) by the original amount and multiply by 100.

You can also check the answer to part (a) by dividing. $15\,600 \div 15\,000 = 1.04$, indicating a 4% increase.

If you get a two-part question like this, make sure you get an answer to part (a). Even if it is wrong you could still get full marks for part (b).

Summary

Percentages are tested in the exam most years. The questions appear in Paper 2 and are among the easier ones in that paper. Questions are likely to be similar to those shown in this section. It is important that you practise extra examples at home to gain confidence and guarantee yourself some vital marks.

Chapter 4

VOLUME

In this outcome, you are expected to be able to find the volumes of spheres, cones and prisms, including cylinders. Rounding to a given number of significant figures is usually tested here. This topic appears in the calculator paper each year, although there may be a non-calculator question as well. We shall explore that later.

Questions are usually worth between 5 and 7 marks, and you will have the advantage of being able to use the formulae sheet. On this you will find :

Key Points

Volume of a sphere: Volume $= \dfrac{4}{3}\pi r^3$

Volume of a cone: Volume $= \dfrac{1}{3}\pi r^2 h$

Volume of a cylinder: Volume $= \pi r^2 h$.

Even if you know a formula, you should check it on the formulae sheet. You would be surprised how many students copy formulae down *incorrectly* from the formulae sheet, so double check each time you use one!

Formulae to Learn

There are some formulae **not** given on the formula sheet which may occur in the exam. You will have met them many times. Make sure you have memorised them.

Area of a square $= l^2$; Area of a rectangle $= lb$; Area of a circle $= \pi r^2$;

Volume of a cube $= l^3$; Volume of a cuboid $= lbh$; Volume of a prism $= Ah$.

The last of these, the volume of a prism, is very important. Remember that A is the area of the base, and h is the height of the prism. We shall do an example on this later.

Significant Figures

This topic is usually tested in a question on volume. You **must** be able to round an answer to a given number of significant figures. For example,

rounding 2826 to two significant figures $\Rightarrow$ 2800

rounding 17 157·284 68 to three significant figures $\Rightarrow$ 17 200.

If you are unsure of how to obtain the answers above, find out before the exam. Remember, too, that if a question asks for a rounded answer, do not round until the end of the question.

The value of π

I would always recommend that you use the π button on your calculator as it is one thing less to memorise; however, if you use the approximate value, $\pi = 3\cdot14$, it will lead to slight variations in the answers, but you will not lose any marks.

If a volume question appears in the Non-calculator paper, you will be told to use $\pi = 3\cdot14$.

We shall now work through a variety of the types of questions you can expect to find in the exam, starting with a straightforward use of one of the formulae.

Example   A sphere has diameter 13·5 centimetres.
Calculate its volume.
Give your answer correct to two significant figures.

(Solution)   Look up the formula. Volume $= \dfrac{4}{3}\pi r^3$.

The formula uses r, the radius, so halve the diameter. Hence $r = 13\cdot5 \div 2 = 6\cdot75$.

Now substitute into the formula; Volume $= \dfrac{4}{3} \times \pi \times 6\cdot75^3$.

Use power/index key on calculator, y^x, or $4 \div 3 \times \pi \times 6\cdot75 \times 6\cdot75 \times 6\cdot75$.

Hence   Volume $= 1288\cdot249337$ (write this down in full)
            $= 1300\,\text{cm}^3$ (to 2 significant figures).

This is worth three marks, for the substitution, calculation and rounding. Always watch out if you are asked for a rounded answer in a volume question. Many students forget to round and lose a mark. You are not normally penalised for missing out units, but it is better to put them in.

The next technique we shall look at is one in which you are given the volume and asked to work back to find the height or the radius. In this type of question you will have to **divide** on your way to the answer. If you are working back to find the **radius** of a cone or cylinder you will have to take the **square root** at the end because both formulae contain r^2. It is unlikely that you would be asked to work back to find the radius of a sphere because at the end you would have to take the cube root, $\sqrt[3]{\ }$, because the formula contains r^3, and this is more difficult.

Example   A cone has a volume of 1092 cubic centimetres.
Its height is 18 centimetres. Find its radius.

(Solution)   Look up the formula. Volume $= \dfrac{1}{3}\pi r^2 h$.

Substitute into the formula: $1092 = \dfrac{1}{3} \times \pi \times r^2 \times 18$.

Divide: $r^2 = \dfrac{1092}{\left(\dfrac{1}{3} \times \pi \times 18\right)} = \dfrac{1092}{18\cdot84955592} = 57\cdot93$.

To find r take the square root: $r = \sqrt{57\cdot93}$

                $= 7\cdot6\,\text{cm}$ (to two significant figures).

This would usually be worth three marks, two of them for substituting the values for the volume and the height, and one for the calculations.

Read through this example carefully and make sure you understand it all.

Next we shall look at longer volume questions. The main types are those in which volumes have to be added or subtracted; two-part questions involving 'working back', and prism questions, usually involving cylinders.

Example

A boiler is in the shape of a cylinder with hemispherical ends as shown in the diagram. The total length of the boiler is 2·2 metres and the length of the cylinder is 1·5 metres.

Find the volume of the boiler.

Give your answer in cubic metres correct to two significant figures.

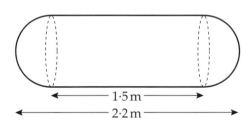

Hints and Tips

You cannot do this unless you know the radius of the cylinder and hemispheres. By subtracting 1·5 from 2·2 you find the width of the two hemispheres, i.e. 0·7 metres. This is therefore the diameter of the sphere formed from the two hemispheres, and the radius is $0·7 \div 2 = 0·35$ metres.

(Solution) Volume of cylinder $= \pi r^2 h$

$$= \pi \times 0·35^2 \times 1·5$$

$$= 0·57726765$$

Volume of two hemispheres $=$ Volume of sphere

$$= \frac{4}{3}\pi r^3$$

$$= \frac{4}{3} \times \pi \times 0·35^3$$

$$= 0·17959438.$$

Total volume of boiler $=$ volume of cylinder $+$ volume of sphere

$$= 0·5773 + 0·1796$$

$$= 0·7569$$

$$= 0·76 \text{ m}^3 \text{ (to two significant figures)}.$$

This question would probably be worth five marks, since it requires finding the radius of the boiler, using the correct strategy (adding a cylinder and sphere), making correct substitutions, and performing correct calculations and rounding.

A good point about these questions is that even if you make a numerical error, it is still possible to get most of the marks as long as you keep the correct method going.

The shape in the last Example is sometimes called a *composite* shape, as it is made up of two different shapes. We find its total volume by adding. We can also have questions where we have to subtract volumes.

Example

The cross-section of a cylindrical metal pipe is shown in the diagram below. The outside diameter is 36 centimetres and the inside diameter is 30 centimetres.

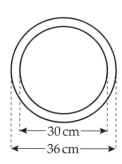

80 cm

←—30 cm—→
←——36 cm——→

The pipe is 80 centimetres long.
Calculate the volume of metal in the pipe.
Give your answer correct to three significant figures.

You can either think of this as the difference between the volume of two cylinders (recommended, as the formula is on your sheet) or as the volume of a prism.

(Solution)

Method 1

Outside cylinder: $r = 18$
Volume $= \pi r^2 h$
$\qquad = \pi \times 18^2 \times 80$
$\qquad = 81\,430$

Inside cylinder: $r = 15$
Volume $= \pi r^2 h$
$\qquad = \pi \times 15^2 \times 80$
$\qquad = 56\,549$
Volume of pipe $= 81\,430 - 56\,549$
$\qquad\qquad = 24\,881$
$\qquad\qquad = 24\,900$ cm^3 (rounded)

Method 2

Outside circle: $r = 18$
Area $= \pi r^2 = \pi \times 18^2$
$\qquad\quad = 1018$

Inside circle: $r = 15$
Area $= \pi r^2 = \pi \times 15^2$
$\qquad\quad = 707$
Area of cross-section $= 1018 - 707$
$\qquad\qquad\qquad = 311$

Volume of pipe $= Ah$
$\qquad\qquad = 311 \times 80$
$\qquad\qquad = 24\,880$
$\qquad\qquad = 24\,900$ cm^3 (rounded)

Answer $= 24\,900$ cm^2 (to 3 significant figures).

The next Example will be a two-part question. This would be worth seven marks, four for part (a) and three for part (b).

Example

Jonathan's toolbox for small tools is in the shape of a prism.

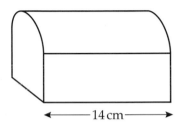

⟵——14 cm——⟶

The length of the toolbox is 14 centimetres.

The cross-section of the toolbox is in the shape of a rectangle and a semi-circle with sizes as shown.

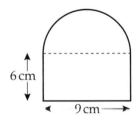

6 cm

◄— 9 cm —►

a)   Find the volume of the toolbox.
     Give your answer correct to two significant figures.

Jonathan buys a new tool container with the **same volume**.

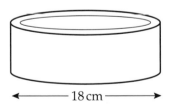

⟵——18 cm——⟶

It is in the shape of a cylinder with diameter 18 centimetres.

b)   Find the height of this toolbox.

Again, there are two approaches to part (a). You can use the formula for the volume of a prism, $V = Ah$. To do this you must find the area of the cross-section, A, by adding the areas of the rectangle and semi-circle, then multiply your answer by h. Be careful, however, because h in this case would be 14, the length of the toolbox. You can try this method and compare your answer with that for the solution given.

Example continued ➣

Example *continued*

(Solution) In this method, for part (a), we will think of the prism as being a cuboid plus a half-cylinder.

a) Volume of prism = volume of half-cylinder + volume of cuboid

$$= \frac{1}{2}\pi r^2 h + lbh$$

$$= 0.5 \times \pi \times 4.5^2 \times 14 \ + \ 14 \times 9 \times 6$$

$$= 1201.3$$

$$= 1200 \text{ cm}^3 \text{ (to 2 significant figures).}$$

Remember

Remember to halve the diameter, 9 cm, to get the radius, 4.5 cm.

Remember to halve the volume of a cylinder. This is often forgotten by students. I have used 0.5 instead of $\frac{1}{2}$ to simplify the calculation.

Remember to write down the un-rounded answer, 1201.3, before rounding.

Note also that there is no need to calculate the volume of the half-cylinder and the volume of the cuboid separately. Using a scientific calculator, $0.5 \times \pi \times 4.5^2 \times 14 + 14 \times 9 \times 6$ can be done in a single calculation.

b) Volume of cylinder = $\pi r^2 h$

Hence $1201.3 = \pi \times 9^2 \times h.$

Hence $h = \dfrac{1201.3}{(\pi \times 9^2)}$

$\Rightarrow$ $h = \dfrac{1201.3}{254.5}$

$\Rightarrow$ $h = 4.7$ cm.

Remember, again, to halve the diameter, 18 cm, to get the radius, 9 cm.

It would be acceptable to use the rounded answer 1200 cm^3 (from part (a)) for the volume rather than 1201.3 cm^3. Slight variations in answers are accepted.

We have now looked at the types of Volume questions which could appear in Paper 2. The fact that you have the formulae sheet and are permitted to use a calculator for the calculations should enable you to get good marks in this topic, as long as you *concentrate* and are *careful*.

To finish, we shall investigate a non-calculator question on Volume. This could appear in Paper 1.

Example

The diagram below shows a cone.

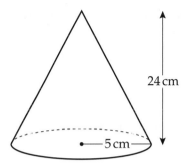

The height is 24 centimetres and the radius of the base 5 centimetres.

Calculate the Volume of the cone.

Take $\pi = 3{\cdot}14$.

(Solution) Volume of cone $= \dfrac{1}{3}\pi r^2 h$

$$= \dfrac{1}{3} \times 3{\cdot}14 \times 5^2 \times 24$$

$$= 628 \text{ cm}^3.$$

This would be worth two marks: one for correct substitution, and one for the correct calculation.

It is fairly easy to work out the answer here without a calculator if you do the operations in the most efficient order. Firstly, get rid of the fraction by dividing 3 into 24, giving 8. The calculation then becomes $3{\cdot}14 \times 5^2 \times 8$. Secondly, work out $5^2 \times 8 = 200$. Finally, $3{\cdot}14 \times 200 = 314 \times 2 = 628$.

THE STRAIGHT LINE

This outcome is usually tested in the Non-calculator paper and is worth between three and five marks.

We shall start by looking at how to draw a line when you are given its equation.

Drawing a Straight Line

Example

Draw the straight line with equation $y = 2x + 1$.

(Solution) Choose some easy values for x, at least three of them, and substitute them into the equation to find the corresponding values for y. I have chosen $x = 0$, 2 and 4. If you substitute these values into $y = 2x + 1$, you will get $y = 1$, 5 and 9. Check these answers. This information is then put into a table.

x	0	2	4
y	1	5	9

Now simply plot the points $(0, 1)$, $(2, 5)$ and $(4, 9)$ and join them. The line can be extended in either direction.

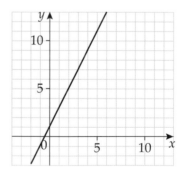

It can be more difficult to make up a table if the equation is given in a slightly different form. For example, consider the straight line with equation $x + 3y = 10$.

If you choose $x = 0$, you get $0 + 3y = 10 \Rightarrow 3y = 10 \Rightarrow y = 3\frac{1}{3}$.

Although the point $(0, 3\frac{1}{3})$ does lie on the line $x + 3y = 10$, it is unsuitable for plotting. You should ignore this and find values of x which give whole number answers for y.

Check that values of 1, 4 and 7 for x lead to values of 3, 2 and 1 for y, therefore the line $x + 3y = 10$ passes through the points $(1, 3)$, $(4, 2)$ and $(7, 1)$.

Crossing the x- and y-axes

You could be asked to find where a line crosses the x-axis or the y-axis.

Remember

Where a line crosses the y-**axis**, $x = 0$.
Where a line crosses the x-**axis**, $y = 0$.

Example

Find the coordinates of the point where the straight line $4y = 10 - 2x$ crosses the x-axis.

(Solution)   The line crosses the x-axis where $y = 0$.
Substitute $y = 0$ into $4y = 10 - 2x \Rightarrow 4 \times 0 = 10 - 2x \Rightarrow 0 = 10 - 2x$
Hence $2x = 10 \Rightarrow x = 5$.
The coordinates are therefore $(5, 0)$

The Equation of a Straight Line

The most likely question on the straight line is to be asked to find the *equation* of a straight line. Worth three marks, this requires knowledge of three things; the gradient, the y-intercept, and the formula $y = mx + c$. We shall now investigate this.

Gradient

The Gradient (or slope) of a straight line can be found by using either of these formulae:

$$\text{Gradient} = \frac{vertical\ height}{horizontal\ distance} \qquad \textbf{or} \qquad \text{Gradient} = \frac{y_2 - y_1}{x_2 - x_1}$$

You should use the one you are familiar with. The letter m is normally used for gradient. Remember that lines sloping up from left to right (/) have a **positive** gradient, while lines sloping down from left to right (\) have a **negative** gradient. Horizontal lines (—) have a **zero** gradient.

y-Intercept

The y-intercept tells us where a straight line cuts the y-axis. If a straight line cuts the y-axis at the point $(0, c)$, then c is the y-intercept.

Remember

The equation $y = mx + c$ is the equation for a straight line with gradient m and y-intercept c. (Sometimes it is written in the form $y = ax + b$, with a the gradient and b the y-intercept.)

Some students may been taught the formula $y - b = m(x - a)$ for finding the equation of a straight line. This formula is helpful if you are going on to study Higher Maths, but it should only be used if you are very familiar with it. The formula $y = mx + c$ is sufficient for any question you could be asked at Intermediate 2.

Example

Find the gradient of the line joining the points $(0, 2)$ and $(6, 4)$.

<u>Method 1</u>   (Plot the points and join them.)

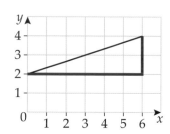

(Gradient is positive)

$$\text{Gradient} = \frac{vertical}{horizontal} = \frac{2}{6} = \frac{1}{3}.$$

<u>Method 2</u>   (Use the gradient formula.) The gradient m of the line joining the points (x_1, y_1) and (x_2, y_2) is given by the formula $m = \dfrac{y_2 - y_1}{x_2 - x_1}$.

(x_1, y_1) and (x_2, y_2) become $(0, 2)$ and $(6, 4)$ so $m = \dfrac{y_2 - y_1}{x_2 - x_1}$

$$= \frac{4 - 2}{6 - 0}$$

$$= \frac{2}{6} = \frac{1}{3}.$$

Whichever method you use, make sure you practise examples from your textbook until you are confident about calculating gradients.

Example

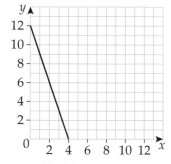

Find the equation of the straight line shown in this diagram.

Example continued >

Example *continued*

(Solution) Find the gradient m; m is negative.

$$m = \frac{vertical}{horizontal} = -\frac{12}{4} = -3.$$

Read off the y-intercept c from graph; $c = 12$.

Substitute m and c into equation $y = mx + c$.

Hence the equation of the straight line is $y = -3x + 12$.

Remember

If you prefer to use the gradient formula, check your answer.
Remember that the slope (\) indicates a negative gradient. Many students forget and lose a mark.

It is **essential** that you show your working in this type of question, setting it out as shown.

Sometimes in an exam the horizontal and vertical scales are different from those shown. Also, the axes may have letters other than x and y. So be alert!

Lines parallel to the axes

Equations such as $x = 4$, $x = -2$, $y = 5$, $y = -3$, etc. are all equations of straight lines and you should be aware of them.

Equations in the form $x =$ 'a number' represent straight lines parallel to the y-axis.

Equations in the form $y =$ 'a number' represent straight lines parallel to the x-axis.

For example, the straight line with equation $x = 4$ passes through the point $(4, 0)$ and is parallel to the y-axis.

Re-arranging the equation $y = mx + c$

Sometimes the equation of a straight line has to be re-arranged into the form $y = mx + c$ in order to find the gradient and y-intercept.

Example Find the gradient and y-intercept of the line with equation $3x + 2y = 10$.

(Solution) Re-arrange the equation into the form $y = mx + c$.

$$3x + 2y = 10 \Rightarrow 2y = -3x + 10 = \Rightarrow y = -\frac{3}{2}x + 5$$

Hence the gradient is $-\frac{3}{2}$ and the y-intercept is 5.

Example

A packet contains 500 grams of cereal when full.
Each day 25 grams of cereal are eaten.

The graph below shows the weight of cereal remaining in the packet (W grams) against time (d days).

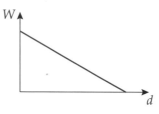

Write down an equation connecting W and d.

(Solution)   After 0 days the weight is 500 grams, i.e. (0, 500) lies on the graph. The packet will last for (500 ÷ 25) days = 20 days, so after 20 days the weight is 0 grams, i.e. (20, 0) lies on the graph.

The graph is therefore as follows

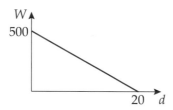

(Gradient m is negative)

Gradient $m = \dfrac{vertical}{horizontal} = -\dfrac{500}{20} = -25.$

y-intercept $c = 500.$

The equation of the straight line is $y = mx + c$
$$\Rightarrow y = -25x + 500.$$

Hence the equation connecting W and d is $W = -25d + 500.$

This is quite tricky, so be aware of this type of problem just in case it occurs.

However, most questions on the straight line are more straightforward, involving gradient, the y-intercept and the formula $y = mx + c$. Once again, prepare properly and you should pick up valuable marks.

ALGEBRA

In this outcome, you must be able to do **two** things to earn all the marks in the exam.

Firstly, multiply out brackets and collect like terms. Secondly, factorise expressions.

Both skills can be learned easily as long as you are prepared to practise them by doing many examples from your textbook. Once you have mastered such skills, you will be able to pick up the marks on offer for this outcome (usually between five and seven).

Multiplying Brackets

'Multiply out the brackets and collect like terms'.

The sentence above is regularly used to introduce questions in the exam. Different teachers use different approaches to multiplying out brackets, although they all come to the same result. Read the example below, try it using the method you are familiar with, and compare your answer with the correct one shown.

Example

Multiply out the brackets and collect like terms:
$$(3p - 5)(2p + 1).$$

(Solution) $(3p - 5)(2p + 1)$
$$= 3p(2p + 1) - 5(2p + 1)$$
$$= 6p^2 + 3p - 10p - 5$$
$$= 6p^2 - 7p - 5.$$

Remember

Remember that a negative sign in front of a bracket changes all the signs inside the bracket. In the above example, the -5 multiplied by $+1$ leads to -5 in the answer.

Example

You may be asked to **square** a bracket such as $(2y - 3)^2$.

(Solution) $(2y - 3)^2$
$$= (2y - 3)(2y - 3)$$
$$= 2y(2y - 3) - 3(2y - 3)$$
$$= 4y^2 - 6y - 6y + 9$$
$$= 4y^2 - 12y + 9.$$

The next example, worth three marks, involves brackets with two terms in the first and three in the second. This type of question occurs *frequently* in exams.

Example   Multiply out the brackets and collect like terms:

$$(4x - 3)(2x^2 - x + 5).$$

(Solution) $(4x - 3)(2x^2 - x + 5)$
$$= 4x(2x^2 - x + 5) - 3(2x^2 - x + 5)$$
$$= 8x^3 - 4x^2 + 20x - 6x^2 + 3x - 15$$
$$= 8x^3 - 10x^2 + 23x - 15.$$

Since you are serious about passing Intermediate 2 Maths, it is important that you practise examples like those above until you are confident that you can get them correct under pressure in an exam. If you cannot do this, it makes factorisation, which we shall look at shortly, impossible.

Before looking at factorisation, here is one more common question on multiplying out brackets.

Example   Multiply out the brackets and collect like terms:

$$(2m + 1)(m - 7) + 5m.$$

(Solution) $(2m + 1)(m - 7) + 5m$
$$= 2m(m - 7) + 1(m - 7) + 5m$$
$$= 2m^2 - 14m + m - 7 + 5m$$
$$= 2m^2 - 8m - 7.$$

Factorisation

Now we shall look at factorisation, the opposite process to multiplying out brackets.

There are three types of factorisation you should know: common factors, difference of two squares, and trinomials. We will look at each in turn.

Common Factors

If you multiply out $5(a + 3)$, you will get $5a + 15$.

If you factorise $5a + 15$, you will get $5(a + 3)$. The common factor is 5.

Check that you can follow the examples below.

Example   Factorise   a) $4z + 20$ b) $pq - pr$ c) $2b^2 - 8b.$

(Solution)   a) $4(z + 5)$ b) $p(q - r)$ c) $2b(b - 4).$

Remember: when you have factorised an algebraic expression, you should always check your answer by multiplying out the brackets. If you are correct, you should go back to what you started with.

Difference of two squares

Suppose you multiply out the brackets $(a + b)(a - b)$.

You will get $a(a - b) + b(a - b) = a^2 - ab + ab - b^2 = a^2 - b^2$.

Therefore if you factorise $a^2 - b^2$, you will get $(a + b)(a \quad b)$.

This type of factorisation is called a *difference of two squares*. It is easy to recognize.

There are **two** terms, separated by a **minus** sign, and both are **squares**. Consider also $x^2 - 25$. It has two terms, separated by a minus sign, and both are squares [$x^2 = (x)^2$ and $25 = 5^2$].

When factorised, $x^2 - 25 = (x)^2 - 5^2 = (x + 5)(x - 5)$.

The terms in both brackets are the same, one has a $+$ sign, the other a $-$ sign. The order of the brackets is unimportant. Note that 5 is the square root of 25.

Example) Factorise   a) $t^2 - 36$ b) $9n^2 - 16$ c) $4p^2 - 81q^2$.

(Solution)   a) $(t + 6)(t - 6)$ b) $(3n + 4)(3n - 4)$ c) $(2p + 9q)(2p - 9q)$.

 Remember you can check by multiplying out the brackets.

Example)

Factorise $2x^2 - 50$.

(Solution)   Although this looks like a difference of two squares, 2 and 50 are not square numbers, so we must use 2 as a common factor first, then we will have a difference of two squares. (Clearly this is a harder example.)

Hence $2x^2 - 50 = 2(x^2 - 25) = 2(x + 5)(x - 5)$.

Trinomials

Earlier we saw that $(3p - 5)(2p + 1) = 6p^2 - 7p - 5$. The expansion $6p^2 - 7p - 5$ is called a *trinomial*. It has three terms. When you factorise a trinomial, your answer will have two brackets, in this case $(3p - 5)(2p + 1)$. (This is worth two marks.)

 There are various ways of going about factorising a trinomial. You must use the method with which you are confident. However, let us examine the factorisation of the expression $3x^2 - 10x - 8$ (very tricky). The first terms in the two brackets must multiply to give $3x^2$, i.e. $3x$ and x. The last terms in the two brackets must multiply to give -8. These could be -2 and 4, or -4 and 2, or -1 and 8, or -8 and 1.

This leads to eight possible combinations; $(3x - 2)(x + 4)$; $(3x + 4)(x - 2)$; $(3x - 4)(x + 2)$; $(3x + 2)(x - 4)$; $(3x - 1)(x + 8)$; $(3x + 8)(x - 1)$; $(3x - 8)(x + 1)$; $(3x + 1)(x - 8)$.

Only one of these eight possibilities will lead to the *correct middle term* of $-10x$. (It is worth noting that you would get 1 out of 2 marks for any of the seven incorrect possibilities.)

To find the correct answer you can multiply out the various brackets and collect like terms.

In fact $(3x + 2)(x - 4) = 3x(x - 4) + 2(x - 4) = 3x^2 - 12x + 2x - 8 = 3x^2 - 10x - 8$ and is the correct answer.

You may think that this will take far too long to do, but it won't if you practise doing lots of trinomials during the year. I would recommend that you practise doing an exercise from a textbook with about forty trinomials (over a period of time), some with x^2, some with $2x^2$, and some with $3x^2$, until you gain in confidence and speed.

If you are planning to go on to try Higher Maths, it is essential that you can factorise trinomials quickly and accurately.

Here are four to try with solutions below.

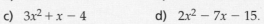

For Practice   Factorise   a) $x^2 - 3x - 10$ b) $2x^2 - 7x + 6$

c) $3x^2 + x - 4$ d) $2x^2 - 7x - 15$.

Answers: a) $(x + 2)(x - 5)$ b) $(2x - 3)(x - 2)$

c) $(3x + 4)(x - 1)$ d) $(2x + 3)(x - 5)$.

Finally in this outcome we shall look at a short algebraic problem worth three marks. There is often one of these in the exam involving multiplying out brackets.

Example   The square and the rectangle below have the **same perimeter**.

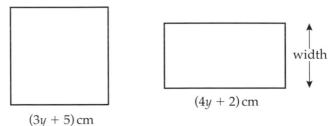

$(3y + 5)\,\text{cm}$

$(4y + 2)\,\text{cm}$

width

Find the width of the rectangle, in centimetres, in terms of y.

Example *continued* ➢

Example *continued*

(Solution)   Perimeter of square $= 4(3y + 5)$
$$= 12y + 20.$$

Perimeter of rectangle $= 2(4y + 2) + (2 \times \text{width})$
$$= 8y + 4 + (2 \times \text{width}).$$

Now perimeter of square = perimeter of rectangle.
Hence $12y + 20 = 8y + 4 + (2 \times \text{width})$
$\Rightarrow$ $2 \times \text{width} = 12y + 20 - (8y + 4)$
$$= 4y + 16.$$

Hence width of rectangle $= (2y + 8)$ *cm*.

Good luck with the algebra questions in the exam. Remember that the only way to become good at multiplying out brackets and factorisation is by lots of practice. This is not going to happen overnight, so you will also require patience.

THE CIRCLE

 The geometry of The Circle is of fundamental importance in maths.

Three main areas are tested in the exam: arcs and sectors (almost guaranteed), symmetry problems involving Pythagoras' Theorem, and angle problems. At least two, and possibly all three, of these areas could be tested, worth between six and ten marks.

The first two areas are likely to appear in the Calculator paper, the third could be in the Non-calculator paper.

Before you start, make sure you have done some revision. You must know how to find the circumference of a circle ($C = \pi d$), and the area of a circle ($A = \pi r^2$). You must be able to use Pythagoras' theorem ($c^2 = a^2 + b^2$), and in order to solve angle problems, you must be able to identify a named angle in a diagram, e.g. $\angle ABC$.

Arcs and Sectors

You could be asked to calculate either the length of an arc **or** the area of a sector.

Remember

Remember that an arc of a circle is part of the **circumference** so you must use the formula $C = \pi d$ in any calculation involving arc length.

Remember that a sector of a circle is an **area** bounded by two radii and an arc so you must use the formula $A = \pi r^2$ in any calculation involving sector area.

In both types of calculation the angle at the centre of the circle must be expressed as a fraction of 360°.

Example)   The diagram below shows a sector of a circle with centre O.

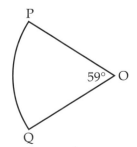

The radius of the circle is 16 centimetres and angle POQ is 59°.

Calculate the length of arc PQ.

Example continued ➤

Example *continued*

(Solution)   Length of arc PQ $= \dfrac{59}{360}$ of circumference $= \dfrac{59}{360} \times \pi d$

$$= \dfrac{59}{360} \times \pi \times 32$$

$$= 16 \cdot 5.$$

Hence length of arc PQ $= 16 \cdot 5$ cm.

Remember here to double the radius, 16, to find the diameter, 32, for the formula. You should use the π button on your calculator, although you would not lose any marks for taking $\pi = 3 \cdot 14$. (This example would be worth three marks.)

Now we shall look at a more difficult practical problem involving the area of a sector.

Example

A windscreen wiper clears an area which is a sector of a circle.

The wiper is 42 centimetres long.

The blade of the wiper is 30 centimetres long.

The wiper swings through an angle of 110° in one sweep.

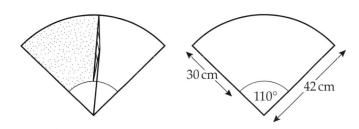

Calculate the area cleared by the wiper in one sweep.

(Solution)   (The area cleared is the difference between the areas of two sectors.)

Area of large sector $= \dfrac{110}{360}$ of $\pi R^2 = \dfrac{110}{360} \times \pi \times 42^2 = 1693 \, \text{cm}^2.$

Radius of small sector $= 42 - 30 = 12 \, \text{cm}.$

Area of small sector $= \dfrac{110}{360}$ of $\pi r^2 = \dfrac{110}{360} \times \pi \times 12^2 = 138 \, \text{cm}^2.$

Area cleared $=$ Area of large sector $-$ Area of small sector

$$= 1693 - 138$$

$$= 1555 \, \text{cm}^2.$$

Check all the working in this example carefully and make sure you understand it. This example is considered quite difficult and is worth four marks (one for $\frac{110}{360}$, one for the correct strategy, one for correct substitutions into the areas of the sectors and one for all calculations done correctly).

You could be given the length of an arc or the area of a sector and asked to find the angle at the centre of the circle. (This would be worth four marks.) To do this you must express either the length of the arc as a fraction of the circumference **or** the area of the sector as a fraction of the area of the circle, and then find this fraction of 360°.

Example

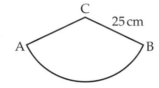

AB is an arc of a circle with centre C and radius 25 centimetres.
The length of arc AB is 54·5 centimetres.
Calculate the size of angle ACB.

(Solution)   Circumference of circle $= \pi d = \pi \times 50 = 157$ cm.

Hence angle $ACB = \dfrac{54 \cdot 5}{157} \times 360 = 125°$.

Again, check carefully, make sure you understand the working, and make sure you practise each type of example (arc length, sector area, finding the angle at the centre) until you are confident.

Symmetry in a Circle

Remember that a line from the centre of a circle perpendicular to a chord bisects the chord. (See the following diagram.)

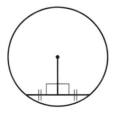

This fact can often help us solve problems in a circle involving the calculations of lengths. Such problems usually depend on the use of Pythagoras' Theorem, although that might not be obvious immediately. Look at the following example.

Example

A tanker has a circular cross-section.
It is partly filled with oil.

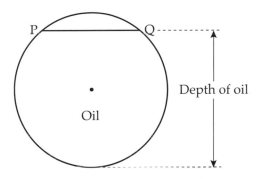

The radius of the circle is 3·2 metres.
The width of the surface of the oil, represented by PQ in the diagram, is 2·4 metres.
Calculate the depth of the oil in the tanker.

◆ This type of calculation (worth four marks) would appear in Paper 2, so you may use a calculator.

◆ As it will be solved using Pythagoras' Theorem, you must create a right-angled triangle. It is essential that you show your triangle as part of your working.

◆ You must mark the centre of the circle with a dot and draw a perpendicular line from the centre to the chord PQ.

◆ Because of symmetry, the perpendicular line will bisect the chord PQ. You can then complete a right-angled triangle by joining the centre to P (or Q).

◆ Now you can use Pythagoras' Theorem to solve the problem.

(Solution)

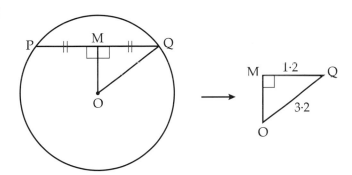

Since PQ = 2·4 m, MQ = 1·2 m.

Using Pythagoras' Theorem, $OQ^2 = OM^2 + MQ^2$

$\Rightarrow \quad 3·2^2 = OM^2 + 1·2^2$

$\Rightarrow \quad OM^2 = 3·2^2 - 1·2^2$

$\quad = 8·8$

$\Rightarrow \quad OM = \sqrt{8·8}$

$\quad = 2·97$ m.

so depth of oil $= 2·97 +$ radius

$\quad = 2·97 + 3·2$

$\quad = 6·17$ m.

In the example, no mention is made of rounding your answer. In this case, you can round your answer to any reasonable degree; 6·17 or 6·2 would be fine.

Problems Involving Angles

We shall now investigate problems in which you are asked to find the size, in degrees, of an angle in a diagram, using the properties of angles in a circle.

Make sure you know which angle you are being asked to find! A surprising number of students cannot identify a named angle such as ∠PQR in a diagram. In fact, an angle named ∠PQR would have its vertex (or corner) at Q, and lie between the arms PQ and RQ. It could also be named ∠RQP.

Remember

Check that you know the following facts about angles.

a) A straight angle $= 180°$.
b) The sum of the three angles in a triangle $= 180°$.
c) Vertically opposite angles are equal.
d) For parallel lines, corresponding angles are equal ('F-shape' angles).
e) For parallel lines, alternate angles are equal ('Z-shape' angles).

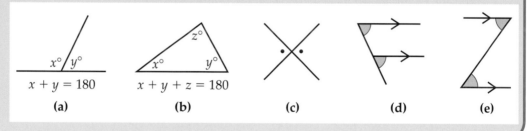

$x + y = 180$ $x + y + z = 180$

    (a)   **(b)**   **(c)**   **(d)**   **(e)**

Tangents

A *tangent* is a straight line which touches a circle at one point only. This point is called the point of contact. The tangent to a circle is perpendicular to the radius at its point of contact.

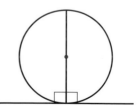

The Angle in a Semi-circle

If PQ is a diameter of a circle and R lies on the circumference, then $\angle PRQ = 90°$.

We say that 'an angle in a semi-circle is a right angle'.

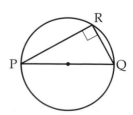

Isosceles Triangles

If, in a circle diagram, a triangle is drawn with two of its sides being radii, then the triangle is isosceles – having two equal sides and two equal angles. Watch out for these triangles as they are often the key to solving angle problems.

In the diagram below, if O is the centre of the circle and B and C lie on the circumference, then triangle OBC is isosceles. Also OB = OC, and $\angle$OBC = $\angle$OCB.

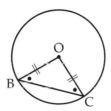

Using this information, you should be able to tackle the type of problems you will meet in the exam. (They are usually worth three marks and appear in Paper 1.)

Example

In this diagram, D, E, and F are points on the circumference of a circle with centre O.

DG is a tangent to the circle, and angle FDG = 36°.

Calculate the size of angle DEF.

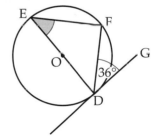

 I have shaded angle DEF in the diagram to help you. It would not be shaded in the exam. Make sure you could identify it if it was not shaded.

(Solution) $\angle$ODG = 90° (angle between tangent and radius).
Hence $\angle$ODF = (90 − 36)° = 54°.
Also $\angle$DFE = 90° (angle in a semi-circle).
Hence $\angle$DEF = (180 − 54 − 90)° = 36° (sum of angles in a triangle = 180°).

Hints and *Tips*

Many students have difficulty setting out their working in this type of question. I would **strongly recommend** that you copy the circle diagram into your exam booklet and fill in the sizes of any angles you calculate on the diagram. You will gain marks for angles you fill in correctly.

Example

MN is a tangent to the circle, centre O,
with point of contact P.
MPN is parallel to QR.
Angle MPQ = 64°.
Calculate the size of angle RQO.

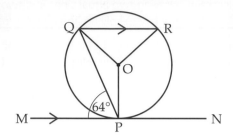

(Solution)   ∠MPO = 90° (angle between tangent and radius)

so   ∠OPQ = (90 − 64)° = 26°,

and   ∠OQP = 26° (triangle OPQ is isosceles as OQ = OP).

Also   ∠PQR = 64° ('Z-shape' as MPN is parallel to QR),

hence   ∠RQO = (64 − 26)° = 38°.

This is quite a difficult example. All the information you were given at the start of the question has been used in the solution. Go through the working carefully. Copy the diagram yourself and fill in the angles to help your understanding. Remember that you can show your working by using a diagram with angle sizes indicated on it.

This completes not only the circle section but also Unit 1.

You can expect 27 marks on Unit 1 in your exam. With a good performance in the questions on this Unit you will be well on the way to the 40 marks you need to pass the exam!

Chapter 8

TRIGONOMETRY

In this outcome, we shall look at the sine, cosine and tangent of angles greater than 90°, and the *triangle formulae* – Area of a triangle, the Sine Rule and the Cosine Rule. We also look at questions involving three-figure bearings. These topics are mainly tested in Paper 2 and advice on how to use your calculator will be provided. Questions from this outcome are worth between five and eight marks. You will be able to use the formulae sheet for most of the questions.

Angles Greater than 90°

You must be able to find the sine, cosine and tangent of angles greater than 90°. This would be tested in Paper 1, so no calculator. Note that sine, cosine and tangent are called *trigonometric ratios*.

To help you, memorise the diagram below and the common rule – **'all, sin, tan, cos'**.

The rule tells you whether each ratio is positive or negative in each of the four quadrants (or quarters) of the diagram. There is also a rule for finding the related **acute** angle, which you will need to complete any question.

2 (90 − 180)	**1** (0 − 90)
SIN(+)	**ALL**(+)
180 − angle	angle
3 (180 − 270)	**4** (270 − 360)
TAN(+)	**COS**(+)
180 + angle	360 − angle

The example below will explain how to use the diagram.

Example

Given that $\sin 30° = 0.5$, what is the value of $\sin 330°$?

(Solution) 330° is in the 4th quadrant. Only **cos** is positive there, so **sin** is negative.
The rule (360 − angle) leads to $(360 − 30)° = 330°$.
Therefore $\sin 330° = \sin (360 − 30)° = − \sin 30° = − 0.5$.

This example would only be worth one mark.

Example

Given that tan 45° = 1, what is the value of tan 225°?

(Solution) 225° is in the 3rd quadrant. **Tan** is positive. The rule (180 + angle) leads to $(180 + 45)° = 225°$.

Therefore $\tan 225° = \tan (180 + 45)° = +\tan 45° = 1$

Avoid the common error of saying $225 = 5 \times 45$, therefore $\tan 225° = 5 \times 1 = 5$.

We shall now look at questions using the triangle formulae. These are worth more marks and would usually appear in Paper 2. Trigonometry questions should be easy to recognise. There are triangles with lengths of sides and sizes of angles marked or to be calculated.

It is very important that you check that your calculator is in DEGREE mode before your exam, and not RAD or GRAD.

Area of a Triangle

If you are asked to find the area of a triangle, look up the formulae sheet for

$$Area = \frac{1}{2} ab \sin C.$$

Example

The sketch shows a triangle ABC.

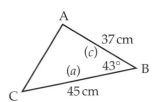

Calculate the area of the triangle.

(Solution) Here, $Area = \frac{1}{2} ac \sin B = 0{\cdot}5 \times 45 \times 37 \times \sin 43° = 568$ cm^2.

This formula requires that you use two sides and the **included** angle, i.e. the angle in between the two sides. Because angle B is given in this problem, the letters in the formula on the sheet should change from $\frac{1}{2} ab \sin C$ to $\frac{1}{2} ac \sin B$ for the Area. Make sure you know how to swap round the letters in a formula to suit the question. Note that the above example is worth two marks.

The Sine Rule and the Cosine Rule

Questions on calculating the length of a side or the size of an angle in a triangle are common. While you can use basic trigonometry (SOH CAH TOA) in a *right-angled triangle*, for all other triangles you need the Sine Rule or the Cosine Rule. Which formula you can use depends on the information you are given.

It will save you time if you can identify which formula to use very quickly.

Here are some instructions to enable you to do decide which formula to use.

Remember

If you are given 3 sides:	Use the 2nd version of the Cosine Rule: $\cos A = \dfrac{b^2 + c^2 - a^2}{2bc}$
If you are given 2 sides and the included angle:	Use the 1st version of the Cosine Rule: $a^2 = b^2 + c^2 - 2bc \cos A$
Anything else:	Use the Sine Rule $\dfrac{a}{\sin A} = \dfrac{b}{\sin B} = \dfrac{c}{\sin C}$

We shall now go through some basic examples on these formulae. These would be worth three marks each, one for knowing to use the correct formula, one for the correct substitution, and one for the correct calculation.

Example

The diagram below shows triangle PQR.

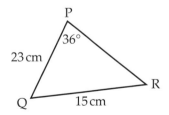

Calculate the size of angle PRQ.

(Solution) $\quad \dfrac{p}{\sin P} = \dfrac{q}{\sin Q} = \dfrac{r}{\sin R} \quad$ (since we are given two sides but *not* the included angle)

$\Rightarrow \dfrac{15}{\sin 36°} = \dfrac{23}{\sin R}$

$\Rightarrow 15 \times \sin R = 23 \times \sin 36° \quad$ (cross-multiply)

$\Rightarrow \sin R = \dfrac{23 \times \sin 36°}{15}$

$\Rightarrow \sin R = 0 \cdot 901$

$\Rightarrow$ angle $R = 64 \cdot 3°$. Hence angle PRQ $= 64 \cdot 3°$.

Example

The diagram below shows a triangle ABC

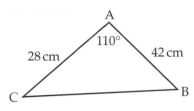

Find the length of side BC.

(Solution) $a^2 = b^2 + c^2 - 2bc \cos A$ (since two sides and included angle are given)

$$\Rightarrow a^2 = 28^2 + 42^2 - 2 \times 28 \times 42 \times \cos 110°$$
$$= 3352$$
$$\Rightarrow a = \sqrt{3352}$$
$$= 57·9 \text{ cm. Hence BC} = 57·9 \text{ cm.}$$

ⓘ When calculating $28^2 + 42^2 - 2 \times 28 \times 42 \times \cos 110°$, I would strongly recommend that you use your calculator to do it all in one go using the square key, x^2, for 28^2 and 42^2. Some students may be tempted to break it down into parts leading to
$784 + 1764 - 2 \times 28 \times 42 \times \cos 110°$, etc.   This can lead to complications and mistakes, and there is only one 'calculation mark' anyway. Try it as I have suggested. It is much easier.

Example

The diagram below shows triangle DEF.

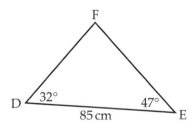

Calculate the length of side EF.

(Solution)   When you know two angles in a triangle, you should calculate the third.
$$\angle DFE = (180 - 32 - 47)° = 101°$$

Now use the Sine Rule $\dfrac{d}{\sin D} = \dfrac{e}{\sin E} = \dfrac{f}{\sin F}$ (since only one side is given).

$$\Rightarrow \frac{d}{\sin 32°} = \frac{85}{\sin 101°}$$

$$\Rightarrow d \times \sin 101° = 85 \times \sin 32°$$

$$\Rightarrow d = \frac{85 \times \sin 32°}{\sin 101°}$$

$$= 45·9 \text{ cm. Hence EF} = 45·9 \text{ cm.}$$

Example

The diagram below shows triangle XYZ.

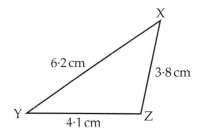

Calculate the size of angle XZY.

(Solution) $\cos Z = \dfrac{x^2 + y^2 - z^2}{2xy}$ (since three sides are given)

$\Rightarrow \cos Z = \dfrac{4\cdot1^2 + 3\cdot8^2 - 6\cdot2^2}{2 \times 4\cdot1 \times 3\cdot8}$

$\qquad = \dfrac{-7\cdot19}{31\cdot16}$ (work out top and bottom separately)

$\qquad = -0\cdot231$ (by division)

$\Rightarrow \qquad Z = 103\cdot4°$

Angle XZY is 103·4°.

It is important that you check all these examples and follow the working. With practice, you should become good at choosing the correct formula and using it properly. Now for some harder examples.

Example

Mr. MacDonald has an allotment in the shape of a quadrilateral.

It is represented in this diagram.

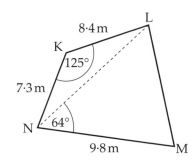

Calculate:
a) The length of the diagonal LN;
b) The area of Mr. MacDonald's allotment.

This example would be worth seven marks, three for part (a) and four for part (b), so it would be vital for you to pick up marks here. You would be told **not** to use a scale drawing. You would get **no** marks if you did.

Example *continued* ➤

Example continued

(Solution) **a)** Use Cosine rule in triangle KLN (two sides and included angle)

$$k^2 = l^2 + n^2 - 2ln \cos K$$
$$= 7\cdot3^2 + 8\cdot4^2 - 2 \times 7\cdot3 \times 8\cdot4 \times \cos 125°$$
$$= 194$$

Hence $k = \sqrt{194}$

Hence LN = 13·9 m.

b) Area of allotment = Area of triangle KLN + Area of triangle LMN.

Area of triangle KLN $= \dfrac{1}{2}ln \sin K = 0\cdot5 \times 7\cdot3 \times 8\cdot4 \times \sin 125°$
$$= 25\cdot1\,\text{m}^2.$$

Area of triangle LMN $= \dfrac{1}{2}lm \sin N = 0\cdot5 \times 9\cdot8 \times 13\cdot9 \times \sin 64°$
$$= 61\cdot2\,\text{m}^2.$$

Hence total Area of allotment $= 25\cdot1 + 61\cdot2 = 86\cdot3\,\text{m}^2$ (approximately).

Check carefully, particularly the substitutions for the areas of the two triangles in part (b). Remember that you need two sides and the included angle for the Area formula.

Example

In order to calculate the height of her school, Angela measures the angle of elevation at two positions A and B as shown in the diagram below.

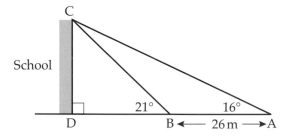

At A, the angle of elevation to C, the top of the school is 16°.
At B, the angle of elevation to C is 21°.
AB is 26 metres.

Calculate the height of Angela's school.

This question would be worth five marks. The extra marks indicate that there is more than just a simple application of the Sine Rule or the Cosine Rule. In fact there are two stages to this type of problem, usually involving either the Sine Rule or the Cosine Rule possibly followed by some basic trigonometry (SOH CAH TOA).
In this case, you should use the Sine Rule in triangle ABC to calculate side BC followed by basic trigonometry in the right-angled triangle CDB.

Example continued ➤

Example *continued*

(Solution) In triangle ABC, $\angle ABC = (180 - 21)° = 159°$

$\angle ACB = (180 - 16 - 159)° = 5°$ (sum of angles in triangle $= 180°$)

Use the Sine Rule: $\dfrac{a}{\sin A} = \dfrac{b}{\sin B} = \dfrac{c}{\sin C} \Rightarrow \dfrac{a}{\sin 16°} = \dfrac{26}{\sin 5°}$

$$\Rightarrow a \times \sin 5° = 26 \times \sin 16°$$

$$\Rightarrow a = \frac{26 \times \sin 16°}{\sin 5°}$$

so BC $= 82 \cdot 2$ m.

Now if h is height of school,

$$\sin 21° = \frac{h}{82 \cdot 2}$$

$$\Rightarrow \quad h = 82 \cdot 2 \times \sin 21°$$

$$= 29 \cdot 5.$$

The height of the school is therefore $29 \cdot 5$ metres.

Three-figure Bearings

Trigonometry questions are sometimes put in a context of three-figure bearings, so you must understand how to tackle such questions.

The three-figure bearing of North is 000°. Three-figure bearings are directions given by angles measured **clockwise from North**, e.g. East has a bearing of 090°.

When you meet such questions in an exam, it is usually easy to pick out a triangle in which to apply the Sine Rule or the Cosine Rule, but the difficulty is in calculating the angle(s) required to use the rule. We shall look at a tricky example on bearings to explain how to proceed. This is worth four marks.

Example In the diagram below a base, a camp, and a village are represented by the points B, C and V respectively.

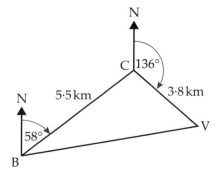

Example *continued* ➢

Example continued

On a cross-country hike, Barry leaves base B and travels 5·5 kilometres on a bearing of 058° to camp C. He then travels 3·8 kilometres from C on a bearing of 136° to village V, and then returns directly to base B.

Calculate the distance Barry travels from the village back to base.

i Clearly triangle BCV is the key to solving this problem. However, we do not know the size of any angles in the triangle. To help you work out the size of angle BCV, you should extend the North–South line at C. By extending lines you can always find F-shape and Z-shape angles to help you. As you can see from this next diagram, $\angle BCV = 102°$. You must show this diagram as part of your working. Now you can use the Cosine Rule (two sides and the included angle).

(Solution) $\angle BCV = (58 + 44)° = 102°.$
Now use the Cosine Rule:

$$c^2 = b^2 + v^2 - 2bv \cos C$$
$$= 3\cdot8^2 + 5\cdot5^2 - 2 \times 3\cdot8 \times 5\cdot5 \times \cos 102°$$
$$= 53\cdot4.$$
$$\Rightarrow c = \sqrt{53\cdot4} = 7\cdot3 \text{ km}.$$

The distance from the village back to the base is therefore 7·3 km.

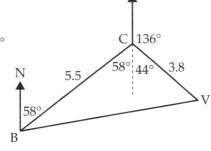

! **Check the working carefully and make sure you understand it. Remember in bearings questions to extend lines, particularly North–South lines. Once you find F-shapes or Z-shapes you can usually calculate the missing angles. Even if you are not sure of the angles, keep going as there are marks available for using the correct triangle formula.**

! **Finally, although it would be unusual, you could get a question on the triangle formulae in Paper 1 – no calculator!**

Example

Here is a triangle RST.

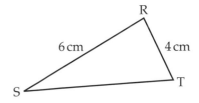

In this triangle RS = 6 cm, RT = 4 cm, and $\sin T = \dfrac{3}{4}$.

Show that $\sin S = \dfrac{1}{2}$.

Example continued ➤

Example continued

(Solution)   Use the Sine Rule $\dfrac{r}{\sin R} = \dfrac{s}{\sin S} = \dfrac{t}{\sin T}$

$\Rightarrow \dfrac{4}{\sin S} = \dfrac{6}{\frac{3}{4}}$

$\Rightarrow 4 \times \dfrac{3}{4} = 6 \times \sin S$

$\Rightarrow 3 = 6 \times \sin S$

$\Rightarrow \sin S = \dfrac{3}{6} = \dfrac{1}{2}.$

Can you do the multiplication of fractions calculation correctly? Check all the working as usual.

Trigonometry is probably the most difficult outcome in Unit 2. However, if you learn when and how to use the triangle formulae, you should be able to pick up lots of marks. So again, plenty of practice is required.

SIMULTANEOUS EQUATIONS

Simultaneous equations occur frequently in mathematical situations. The equations involve two (or sometimes more) unknowns. However, these unknowns have the same value in each equation!

You should be able to solve simultaneous equations such as:

$$5x + 3y = 18$$
$$4x - 2y = 10$$

Try to find the values of x and y.

A question on simultaneous equations (sometimes called a *system of equations*) appears in the exam every year, worth three or four marks for a straightforward example, or even six marks for a more difficult problem.

When solving such equations, make sure that the equations are set up as shown above, with the same variables in line and the numerical term on the right hand side. The recommended method is called *elimination* in which we 'get rid' of one of the variables. You should number the equations (1) and (2) and *scale* the equations so that the coefficients of one of the variables are the same. The coefficients are the numbers in front of the variables (x and y). You can then eliminate that variable by adding or subtracting the equations.

Well done if you found $x = 3$ and $y = 1$ in the above example.

Example

Solve algebraically the system of equations $\quad 3x + 2y = 5$
$\qquad\qquad\qquad\qquad\qquad\qquad\qquad\qquad\quad 2x - 5y = 16.$

(Solution) $\quad 3x + 2y = 5$ **(1)**
$\qquad\qquad 2x - 5y = 16$ **(2)**

$\qquad$ **(1)** $\times 5$: $\;15x + 10y = 25$ **(3)**
$\qquad$ **(2)** $\times 2$: $\quad 4x - 10y = 32$ **(4)**

$\qquad$ **(3)** $+$ **(4)**: $\qquad\;\; 19x = 57$
$\qquad \Rightarrow \qquad\qquad\quad\; x = 57 \div 19$
$\qquad \Rightarrow \qquad\qquad\quad\; x = 3.$

Example *continued* $\succ$

Example continued

Now substitute $x = 3$ into equation (**1**) $3x + 2y = 5$.

$$
\begin{aligned}
\text{Hence} \quad & 3 \times 3 + 2y = 5 \\
\Rightarrow \quad & 9 + 2y = 5 \\
\Rightarrow \quad & 2y = 5 - 9 \\
\Rightarrow \quad & 2y = -4 \\
\Rightarrow \quad & y = -2.
\end{aligned}
$$

Solution is therefore $x = 3$, $y = -2$.

The word 'algebraically' in the question tells you to solve the equations as shown above rather than by a *graphical method*. The process where we multiply the equations to make the y-coefficients the same, i.e. 10, is called *scaling*. Note that we **add** equations (**3**) and (**4**) because the signs are opposite, $+10$ and -10, thus eliminating y. When the signs are the same, both $+$ or both $-$, we subtract the equations. You may have been taught to multiply one equation by a negative number and then add. That is fine if you are comfortable doing it that way.

Example

Solve algebraically the system of equations $\quad 4a + 3b = 7$
$\qquad\qquad\qquad\qquad\qquad\qquad\qquad\qquad\qquad 2a + 7b = 31.$

$$
\begin{aligned}
(\text{Solution}) \quad & 4a + 3b = 7 & \mathbf{(1)} \\
& 2a + 7b = 31 & \mathbf{(2)} \\[6pt]
& \mathbf{(1)} \times 1: \quad 4a + 3b = 7 & \mathbf{(3)} \\
& \mathbf{(2)} \times 2: \quad 4a + 14b = 62 & \mathbf{(4)} \\[6pt]
& \mathbf{(4)} - \mathbf{(3)}: \quad 11b = 55 & \\
& \Rightarrow \qquad\qquad b = 5.
\end{aligned}
$$

Now substitute $b = 5$ into equation (**1**) $4a + 3b = 7$.

$$
\begin{aligned}
\text{Hence} \quad & 4a + 3 \times 5 = 7 \\
\Rightarrow \quad & 4a + 15 = 7 \\
\Rightarrow \quad & 4a = 7 - 15 \\
\Rightarrow \quad & 4a = -8 \\
\Rightarrow \quad & a = -2.
\end{aligned}
$$

The solution is therefore $a = -2$, $b = 5$.

We eliminated the first variable, a, because only one equation had to be multiplied. (For practice, try the same question by eliminating b.)

When subtracting equations, look for the easier order. Here (**4**) – (**3**) avoids negative numbers.

Remember, too, that you can always check your answers by substituting both values in equation (**2**). Thus $2a + 7b = (2 \times -2) + (7 \times 5) = -4 + 35 = 31$ (as required).

Example

Find the point of intersection of the straight lines with equations $x + 2y = 6$ and $3x - 5y = 7$.

(Solution) This type of question can be done by making up tables of values for x and y, drawing both lines and reading off where they intersect. However, it is much simpler to solve simultaneous equations.

$$x + 2y = 6 \qquad \textbf{(1)}$$
$$3x - 5y = 7 \qquad \textbf{(2)}$$

$$\textbf{(1)} \times 3: \quad 3x + 6y = 18 \qquad \textbf{(3)}$$
$$\textbf{(2)} \times 1: \quad 3x - 5y = 7 \qquad \textbf{(4)}$$

$$\textbf{(3)} - \textbf{(4)}: \qquad 11y = 11 \qquad (\text{Note } 6 - (-5) = 11)$$
$$\Rightarrow \qquad \qquad y = 1.$$

Now substitute $y = 1$ into equation **(1)**
$$x + 2y = 6$$
$$\Rightarrow \qquad x + 2 \times 1 = 6$$
$$\Rightarrow \qquad x + 2 = 6$$
$$\Rightarrow \qquad x = 4.$$

Hence the point of intersection is (4, 1).

Remember to give the answer as coordinates here or you will lose a mark.

It is quite common for a six-mark problem to appear in the exam. We shall look at this type of question next. In it, part (a) is worth one mark; part (b), one mark; and part (c), four marks. You are allowed a calculator in this example.

Example

a) Alice bought a bouquet of 6 daffodils and 4 tulips from a florist.
Let d pence be the cost of a daffodil, and t pence be the cost of a tulip.
The cost of the bouquet was £5·30.
Write down an equation in d and t which satisfies the above condition.

b) David bought a bouquet of 5 daffodils and 7 tulips from the same florist.
He paid £6·80 for the bouquet.
Write down a second equation in d and t which satisfies this condition.

c) Find the cost of one daffodil and the cost of one tulip.

Example *continued* ➢

Example *continued*

(Solution) a) $6d + 4t = 530$

b) $5d + 7t = 680$

c) $6d + 4t = 530$ **(1)**

$5d + 7t = 680$ **(2)**

(1) ×7: $42d + 28t = 3710$ **(3)**

(2) ×4: $20d + 28t = 2720$ **(4)**

(3) − **(4)**: $22d = 990$

$\Rightarrow$ $d = 990 \div 22 = 45.$

Now substitute $d = 45$ into equation **(1)** $6d + 4t = 530$

$\Rightarrow$ $6 \times 45 + 4t = 530$

$\Rightarrow$ $270 + 4t = 530$

$\Rightarrow$ $4t = 530 - 270$

$\Rightarrow$ $4t = 260$

$\Rightarrow$ $t = 65.$

Hence a daffodil costs 45 pence and a tulip costs 65 pence.

There is a *communication* mark in this type of question for writing your answer in words. Don't forget or you will lose a mark.

Remember that you can *check* your answer. Show your working clearly because the correct answer with no working gets no marks.

In this type of problem, where the answers must be positive numbers, you should subtract the scaled equations **(3)** and **(4)**. If the problem involves units of measurement such as centimetres, then you must include these units of measurement in the final sentence.

It is easy to make a mistake in your working. If your answers include strange fractions you have probably made a mistake, so check your working. If you cannot find the mistake, don't score out what you have done. You could still get five out of six marks even with the wrong answer, if your strategy and communication are correct.

However, simultaneous equations are easy to solve if you have practised enough examples. So there are no excuses for losing marks here.

Chapter 10

GRAPHS, CHARTS AND TABLES

Questions on Graphs, Charts and Tables are worth between six and eight marks in the exam. You should aim to pick up full marks here, as this work is very straightforward. You should already be able to interpret data from bar graphs, line graphs, stem and leaf diagrams and scattergraphs, as well as being able to construct them. We shall look in detail at some other types of data displays in this section.

Pie Charts

You may be asked to construct a pie chart, so be prepared. Always take compasses, a protractor and a sharp pencil into your exam. Constructing a pie chart is based on the fact that there are 360° in a circle. Study this example.

Example

A sample of students was asked which flavour of crisps they preferred.
Their responses are shown below.

Flavour	Frequency
Ready Salted	20
Cheese and Onion	10
Salt and Vinegar	30
Prawn Cocktail	20

Construct a pie chart to illustrate this information.

You must show your working. You can use a calculator. This is worth 3 marks.

(Solution)   Total number of students $= 20 + 10 + 30 + 20 = 80$.
(Each group is now expressed as a fraction of the total and then converted into a sector of a circle.)

Ready Salted $\dfrac{20}{80} \times 360° = 90°$;

Cheese and Onion $\dfrac{10}{80} \times 360° = 45°$;

Salt and Vinegar $\dfrac{30}{80} \times 360° = 135°$;

Prawn Cocktail $\dfrac{20}{80} \times 360° = 90°$.

We now draw a circle and draw the four sectors, by measuring angles of 90°, 45°, 135° and 90° with a protractor.

Remember that you must label your diagram. The *order* of the sectors is unimportant.

Cumulative Frequency

This is the total frequency up to a particular item in a data set. It can be thought of as a 'running total' within the data set.

Example)

A group of children was asked how many coins each had. The results are as follows:

5 4 2 0 5 4 5 2 1 0 3 2
1 2 5 2 3 4 4 3 2 1 4 0.

Construct a frequency table for the above data and add a cumulative frequency column.

(Solution) Use tally marks to help with the frequency, and add the frequencies as you go along to get the cumulative frequency.

Number of coins		Frequency	Cumulative frequency
0					3	3
1					3	6
2								6	12
3					3	15
4							5	20
5						4	24

Always count how many numbers you are given and make sure it is the same as the last entry in the cumulative frequency column.

The Median and Quartiles

Key Words *and* **Definitions**)

The **median** is the **middle** value in an ordered data set. If there are n entries in the data set, the formula $(n + 1) \div 2$ enables you to find the **position** of the median. For example, if you have a set of 15 ordered numbers, by calculating $(15 + 1) \div 2 = 8$, then the 8th number is the median. If you have 16 ordered numbers, then $(16 + 1) \div 2 = 8 \cdot 5$, meaning that the median is mid-way between the 8th and 9th numbers.

The median splits a data set into two halves. The **lower quartile** is the median of the lower half and the **upper quartile** is the median of the upper half.

We use the symbols Q_1, Q_2 and Q_3 for the lower quartile, the median and the upper quartile respectively.

To find a median that is mid-way between two numbers, we calculate the mean value of the two numbers.

Example

For the data set   7, 9, 5, 10, 14, 7, 4, 11, 14, 6, 8, 7, 9, 6, 5, 19, 14

Find   a) the median;
           b) the lower quartile;
           c) the upper quartile.

(Solution)   Put the numbers in ascending order. Always count the original list and the ordered list to make sure you have not missed any out. Here $n = 17$.

$$4 \ 5 \ 5 \ 6 \mid 6 \ 7 \ 7 \ 7 \ \underline{8} \ 9 \ 9 \ 10 \ 11 \mid 14 \ 14 \ 14 \ 19$$

a) The median, $Q_2 = 8$ [the 9th number along as $(17 + 1) \div 2 = 9$].
b) The lower quartile, $Q_1 = 6$ (between the 4th and 5th numbers).
c) The upper quartile, $Q_3 = 12 \cdot 5$ (between the 13th and 14th numbers).

(Note that the value $12 \cdot 5$ is calculated as $(11 + 14) \div 2$.)

Example

The frequency table below shows the marks of some students in a test.

Mark	Frequency
4	6
5	9
6	10
7	8
8	15
9	16
10	16

Calculate   a) the median;
             b) the lower quartile;
             c) the upper quartile.

 You must add the frequencies to find the total number of students. The marks are already in order, so find the position of the median. Then add the frequencies as in a cumulative frequency column until you come to the median mark.
You could think of every mark written out in a long list starting 4 4 4 4 4 4 5 5 ..., although it would take too long to do the question this way.

Example continued ➤

Example continued

(Solution)   Number of students $= 6 + 9 + 10 + 8 + 15 + 16 + 16 = 80$.
Position of median $= (80 + 1) \div 2 = 40.5$ (between 40th and 41st marks).
There are $(6 + 9 + 10 + 8) = 33$ marks of 7 or below.
There are $(6 + 9 + 10 + 8 + 15) = 48$ marks of 8 or below.
Therefore the 40th and 41st marks must be 8. This is the median.
Similarly check that the lower quartile $= 6$ and the upper quartile $= 9$.
Hence **(a)** $Q_2 = 8$, **(b)** $Q_1 = 6$, **(c)** $Q_3 = 9$.

Dotplots

Dotplots provide a straightforward means of illustrating the shape of a distribution of data.

Example

Alex counts the number of paperclips in a sample of boxes.

90 87 88 90 90 91 93 88 89 88 92 86 90 89 91
89 91 92 89 88 93 92 90 91 89 90 90 91 92 90

Illustrate this data in a dotplot.

(Solution)

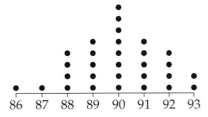

number of paperclips

Dotplots are easy to draw. Set them out as shown, with a label. The above dotplot has a symmetrical distribution. If you are asked to find the median and quartiles from a dotplot, remember that the data is already ordered, count the number of dots, find the position of the median and count along the dots until you come to it.

You should be aware of the following distributions of data:

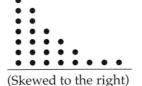
(Skewed to the right)

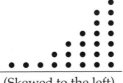
(Skewed to the left)

If the data 'tails' to the right, it is said to be *positively skewed*, or skewed to the right. If it 'tails' to the left, it is *negatively skewed*, or skewed to the left. Skewness occurs when the mean and the median have different values. In a right skew, the mean has a higher value than the median.

Boxplots

There is a question involving boxplots in the exam most years. There are three marks for finding the median, the lower quartile and the upper quartile, and a further two marks for drawing a boxplot. There is often a further question involving comparing two boxplots, so this is also very important for earning marks. You need a **five-figure summary** to draw a boxplot.

First, find the median, then the quartiles, and, finally, the highest and lowest members of the data set.

Example

Mr. Wong waits for the number 66 bus every morning at 9 o'clock.
He keeps a record, in minutes, of his waiting times over a two-week period.
These times are as follows:

5 8 6 5 10 6 9 3 5 12 20 6 8 13.

a) For the given data, calculate:
 (1) the median;
 (2) the lower quartile;
 (3) the upper quartile.

b) Draw a boxplot to illustrate this data.

 Mrs.Greenberg waits for the number 41 bus every morning at 9 o'clock.
 She keeps a record of her waiting times over the same period. Her results are shown in the boxplot below.

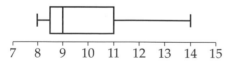

c) By comparing boxplots, make **two** appropriate comments about the waiting times of Mr. Wong and Mrs. Greenberg.

(Solution) a) Order the numbers   3 5 5 $\underline{5}$ 6 6 6 | 8 8 9 $\underline{10}$ 12 13 20.

 Position of Median $= (14 + 1) \div 2 = 7{\cdot}5$ (between 7th and 8th numbers).

 Hence median $= \dfrac{6 + 8}{2} = 7$. The quartiles are obvious from the number line.

 Hence (1) $Q_2 = 7$, (2) $Q_1 = 5$, (3) $Q_3 = 10$.

Example continued ➤

Example continued

b) For the five-figure summary, L (lowest) = 3, H (highest) = 20.

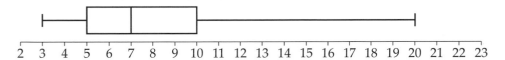

c) Mrs. Greenberg seems to have a longer waiting time. Her median is 9 minutes compared to 7 minutes for Mr. Wong. However, her times are less spread out than his. He has a range of (20 − 3) = 17 minutes, while her range is only (14 − 8) = 6 minutes.

 If you are asked to compare two boxplots, you should comment on the medians, saying which is greater, **and** make a comment about the spread of the boxplots. One way to do this would be to find the range (highest − lowest) for each boxplot and say which is greater. It is always better to write too much than too little.

Example

The number of goals scored by teams in the local football league are recorded in the *stem and leaf diagram* shown below.

Number of goals

```
3 | 5
4 | 0 2 6
5 | 0 1 1 3 5 7 9
6 | 2 2 5 8
7 | 1 3 8
8 | 0 2
```

$n = 20$ 3|5 represents 35 goals

For the given data, calculate:
(1) the median;
(2) the upper quartile;
(3) the lower quartile.

(Solution)   As $n = 20$, the median is in position $(20 + 1) \div 2 = 10.5$, i.e. between the 10th and 11th numbers. The diagram has already ordered the numbers, so the median is between 57 and 59, and is therefore 58.
The lower quartile is the median of the ten numbers in the lower half and is between 50 and 51, and is therefore 50·5.
Similarly, the upper quartile is between 68 and 71 and is therefore 69·5.
Hence (1) $Q_2 = 58$, (2) $Q_1 = 50.5$, (3) $Q_3 = 69.5$.

You should also be able to interpret a *back-to-back* stem and leaf diagram.

At the end of this outcome you should be able to construct and interpret pie charts and dotplots; add a cumulative frequency column to a frequency table; find the median, upper quartile and lower quartile from a data set; draw and compare boxplots.

If you can do all these things accurately, your chances of doing well in the exam are greatly improved.

STATISTICS

We shall look mainly at *measures of spread* in this outcome, with the semi-interquartile range and standard deviation being most important. We shall also look at probability.

This outcome accounts for between five and eight marks, and it, too, is straightforward if you prepare properly.

Semi-interquartile Range

The semi-interquartile range is a measure of the spread of the data in a distribution. The greater the semi-interquartile range, the more spread out the data is. It is found using the formula

$$\text{Semi-interquartile range} = \frac{Q_3 - Q_1}{2}$$

It is half of the difference between the upper and lower quartiles. Memorise the formula as it is not on your formulae sheet.

There is also a formula for the **interquartile range** $= Q_3 - Q_1$, although it is not so common.

Example

The weights of a group of girls, in kilograms, were measured as follows:

$$45 \quad 53 \quad 48 \quad 50 \quad 44 \quad 40 \quad 56 \quad 49 \quad 47 \quad 50 \quad 52 \quad 43 \quad 46.$$

The weights of a group of boys in the same class were also measured in kilograms. The semi-interquartile range for the boys was 4·5.

Make an appropriate comment about the distribution of data in the two groups.

(Solution)   We must calculate the semi-interquartile range for the girls.

We put the numbers in ascending order:

$$40 \quad 43 \quad 44 \ \big| \ 45 \quad 46 \quad 47 \quad \underline{48} \quad 49 \quad 50 \quad 50 \ \big| \ 52 \quad 53 \quad 56.$$

There are 13 numbers, so the median is in $(13 + 1) \div 2 = 7^{\text{th}}$ position.

Median, $Q_2 = 48$. Also from the list of numbers $Q_1 = 44 \cdot 5$, $Q_3 = 51$.

Semi-interquartile range $= \dfrac{Q_3 - Q_1}{2} = \dfrac{51 - 44 \cdot 5}{2} = \dfrac{6 \cdot 5}{2} = 3 \cdot 25$.

As the semi-interquartile range for the girls, 3·25, is less than that for the boys, 4·5, it means that the girls' weights are not spread out as much as the boys' weights.

The semi-interquartile range describes only the spread of the distribution. It does not mean that the girls are lighter than the boys. We can only tell if that is the case by calculating the mean or median of the weights.

Next we shall look at *standard deviation*. This has been tested in every Intermediate 2 exam so far and is obviously very important.

Standard Deviation

Standard Deviation, s, is also a measure of spread. The higher the standard deviation, the more spread out, or varied, the data is. If the standard deviation is small, it means that the data is not spread out very much and that results are consistent. You must bear this in mind when asked to compare two different sets of data based on their standard deviations.

There are two formulae for calculating standard deviation on the formulae sheet. Use the one you are more familiar with.

Example

a) Sarah measures the temperature, in degrees Celsius, at 12 noon every day for a week in July. Her results are as follows:

$$20 \ 25 \ 18 \ 24 \ 22 \ 28 \ 31.$$

For this data calculate the mean and standard deviation.

b) For a week in September, Sarah calculates that the mean temperature is 17°C and the standard deviation is 2·4.

Compare the temperatures in July with those in September.

Part (a) is worth four marks, and part (b) is worth two marks.

(Solution) a) Mean, $\bar{x} = (20 + 25 + 18 + 24 + 22 + 28 + 31) \div 7 = 168 \div 7 = 24\,°C.$

Standard Deviation, s

Method 1		
$s = \sqrt{\dfrac{\sum(x - \bar{x})^2}{n - 1}}$		
x	$(x - \bar{x})$	$(x - \bar{x})^2$
20	−4	16
25	1	1
18	−6	36
24	0	0
22	−2	4
28	4	16
31	7	49
		122

$$s = \sqrt{\frac{122}{7-1}} = \sqrt{\frac{122}{6}} = \sqrt{20\cdot3}$$
$$= 4\cdot5.$$

Method 2	
$s = \sqrt{\dfrac{\sum x^2 - (\sum x)^2/n}{n - 1}}$	
x	x^2
20	400
25	625
18	324
24	576
22	484
28	784
31	961
168	4154

$$s = \sqrt{\frac{4154 - 168^2 \div 7}{7 - 1}} = \sqrt{\frac{122}{6}} = \sqrt{20\cdot3}$$
$$= 4\cdot5.$$

b) The mean is higher in July (24°C) than in September (17°C), so it is warmer in July. The standard deviation is lower in September (2·4) than in July (4·5).

This means that the temperatures in September were less spread out than in July.

Hints and Tips

Check carefully all the working in the formula you have chosen and make sure you know 'what goes where'. Remember that the symbol Σ, sigma, means 'the sum of'.

If you use Method 1, remember that $(x - \bar{x})$ is found by subtracting the mean, $\bar{x}$, from each temperature in turn, and that when you square, the answer is **positive**.

If you use Method 2, calculate $4154 - 168^2 \div 7$ all in one go, leading to 122.

With practice, you should become proficient at calculating standard deviation!

For Practice

Practise by finding the mean and standard deviation of 24, 27, 28, 30, 23 and 30.

Hopefully you will get 27 for the mean and 2·97 for the standard deviation.

Example

The ages (in years) of the players in a netball team are given below:

$$18 \quad 21 \quad 19 \quad 22 \quad 27 \quad 31.$$

a) Calculate the mean and standard deviation of these ages.

b) Write down the mean and standard deviation of the ages 5 years later.

(Solution) a) Mean, $\bar{x} = (18 + 21 + 19 + 22 + 27 + 31) \div 6 = 138 \div 6 = 23$.

Standard Deviation, s

Method 1			**Method 2**	
$s = \sqrt{\dfrac{\Sigma(x - \bar{x})^2}{n - 1}}$			$s = \sqrt{\dfrac{\Sigma x^2 - (\Sigma x)^2/n}{n - 1}}$	
x	$(x - \bar{x})$	$(x - \bar{x})^2$	x	x^2
18	-5	25	18	324
21	-2	4	21	441
19	-4	16	19	361
22	-1	1	22	484
27	4	16	27	729
31	8	64	31	961
		126	138	3300

$$s = \sqrt{\frac{126}{6 - 1}} = \sqrt{\frac{126}{5}} = \sqrt{25 \cdot 2} \qquad\qquad s = \sqrt{\frac{3300 - 138^2 \div 6}{6 - 1}} = \sqrt{\frac{126}{5}} = \sqrt{25 \cdot 2}$$

$$= 5 \cdot 02. \qquad\qquad\qquad\qquad\qquad = 5 \cdot 02.$$

b) 5 years later,

mean, $\bar{x} = 23 + 5 = 28$, standard deviation, $s = 5 \cdot 02$.

Hints and Tips

Some students might add 5 to each number in the original data set and then calculate the mean and standard deviation using the formulae again. They would get the correct answers, but it is not necessary to do so much work. You are asked to **write down** the answers. The new mean is the old mean, $23 + 5 = 28$. By adding 5 to **each** number, the *spread* is unchanged, so the standard deviation stays the same, 5·02.

The previous example can help us to understand how the mean and standard deviation of a data set are affected by changes to the data.

If the same number is **added** to each member of a data set, the mean will increase by that amount, but the standard deviation **will stay the same**.

If each member of a data set is **multiplied** by the same number, both the mean **and** the standard deviation will be multiplied by that number, e.g. if a data set has mean 15 and standard deviation 3, and each member of the data set is doubled, the new mean will be 30 and the new standard deviation will be 6.

Scattergraphs

We looked at how to draw a scattergraph and a best fitting straight line in Chapter 2 (Revision). You could, however, be asked to find the *equation* of a best-fitting line in the exam. If this happens, you must think of the question as 'finding the equation of a straight line'. That is, find the gradient, the y-intercept, and use the formula $y = mx + c$. You may wish to look back at Chapter 5.

If you are asked to estimate a y-value given an x-value, then simply substitute the x-value into the equation of the straight line and do a calculation.

Probability

There is usually a probability question in the exam worth one mark. Remember that probability is a measure of chance between 0 and 1.

If an event is **impossible**, its probability $= 0$.

If an event is **certain**, its probability $= 1$.

All other probabilities lie somewhere in between.

Probability is defined to be:

$$\frac{number\ of\ favourable\ outcomes}{total\ number\ of\ outcomes}$$

STATISTICS

Example

A group of girls were asked their favourite colour. Their results are shown below.

Red 14, Blue 17, Yellow 7, Green 17, Pink 23, Orange 9, Purple 13.

If a girl is chosen at random from the group, what is the probability that her favourite colour is pink?

(Solution)   Total no. of girls $= (14 + 17 + 7 + 17 + 23 + 9 + 13) = 100$.

Hence probability of pink $= \dfrac{23}{100}$.

Hopefully you have found the topics in this outcome to be straightforward. Try to make sure you are very proficient at *standard deviation* as it is worth most marks.

Units 1 and 2 account for 54 of the 80 marks in the Intermediate 2 Exam. Next, we shall look at the optional units, Unit 3 and Unit 4 (Applications of Mathematics).

FURTHER ALGEBRA

Only read Chapters 12, 13 and 14 if you are studying Unit 3. If you are studying Unit 4, the Applications of Mathematics, go straight to Chapter 15.

◆ Unit 3 is considered to be more difficult than Units 1 and 2. It is worth a total of 26 marks in the exam. The work in this Unit will be invaluable if you are planning to continue with Maths at Higher level.

◆ There are several short topics to be studied in this outcome – algebraic fractions, changing the subject of a formula, surds, and indices. All of these topics could appear in Paper 1, so you should not require a calculator. There are usually between 10 and 12 marks on this outcome in the exam.

Algebraic Fractions

We shall start by looking at how to express a fraction in **simplest form**.

If you are asked to simplify $\dfrac{15}{35}$, you should express it as $\dfrac{3 \times 5}{7 \times 5}$ and by cancelling the 5s, arrive at the answer $\dfrac{3}{7}$. You can simplify algebraic fractions in exactly the same way. For example, $\dfrac{ab}{ac} = \dfrac{b}{c}$, by cancelling the a's. Note that you cannot simplify fractions such as $\dfrac{a+b}{a-b}$ because of the $+$ and $-$ signs. If we are asked to simplify $\dfrac{(a+b)^2}{(a+b)(a-b)}$, however, we can cancel an $(a+b)$ bracket from the numerator and denominator of the fraction leading to $\dfrac{a+b}{a-b}$. In some cases, it is necessary to **factorise** the numerator or denominator or both before you can simplify the fraction.

Example

1. Express $\dfrac{p^3 q}{pq^2}$ in its simplest form.

2. a) Factorise $2x + 4$
   b) Factorise $x^2 - 3x - 10$

   c) Hence express $\dfrac{2x + 4}{x^2 - 3x - 10}$ in its simplest form.

Question 1 is worth one mark. Question 2 is worth five marks $(1 + 2 + 2)$.

(Solution) 1 $\dfrac{p^3 q}{pq^2} = \dfrac{p \times p \times p \times q}{p \times q \times q} = \dfrac{\cancel{p} \times p \times p \times \cancel{q}}{\cancel{p} \times q \times \cancel{q}} = \dfrac{p^2}{q}$

2 a) $2x + 4 = 2(x + 2)$ (common factor)

   b) $x^2 - 3x - 10 = (x - 5)(x + 2)$ (trinomial)

   c) $\dfrac{2x + 4}{x^2 - 3x - 10} = \dfrac{2\cancel{(x+2)}}{(x - 5)\cancel{(x+2)}} = \dfrac{2}{x - 5}$

Remember that you may have to factorise the numerator or denominator of a fraction before you can simplify it. This is another reason why factorisation (studied in Chapter 6) is so important. Note also that the word 'hence', in Example 2(c), indicates that you should use the answers you obtained in parts 2(a) and 2(b) of the question to proceed.

The Four Operations for Algebraic Fractions

The four operations are addition, subtraction, multiplication and division. You can expect to be tested on one of these four in your exam, so you must be prepared for all four. Addition and subtraction are very similar, so we shall look at them together.

Addition and Subtraction

Before adding or subtracting fractions, they must be expressed with the same denominator, called the **LCM** (**lowest common multiple**).

Look at a numerical example; $\frac{3}{5} + \frac{1}{4}$. The LCM of 5 and 4 is 20 (5×4), so both fractions must be changed to twentieths. Thus $\frac{3}{5} + \frac{1}{4} = \frac{3 \times 4}{5 \times 4} + \frac{1 \times 5}{4 \times 5} = \frac{12}{20} + \frac{5}{20} = \frac{17}{20}$.

We deal with algebraic fractions in the same way.

Example

a) Express $\dfrac{2}{x+5} + \dfrac{4}{x-2}$ ($x \neq -5, x \neq 2$) as a single fraction in its simplest form.

b) Express $\dfrac{3}{y-1} - \dfrac{4}{y}$ ($y \neq 1, y \neq 0$) as a single fraction in its simplest form.

($x \neq -5$, $x \neq 2$ in part (a) and $y \neq 1$, $y \neq 0$ in part (b) show values which x and y cannot take. Ignore this information when you are doing your working.)

(Solution)   a) $\text{LCM} = (x+5)(x-2)$

Hence $\dfrac{2}{x+5} + \dfrac{4}{x-2} = \dfrac{2(x-2) + 4(x+5)}{(x+5)(x-2)} = \dfrac{2x - 4 + 4x + 20}{(x+5)(x-2)} = \dfrac{6x + 16}{(x+5)(x-2)}$

b) $\text{LCM} = y(y-1)$

Hence $\dfrac{3}{y-1} - \dfrac{4}{y} = \dfrac{3y - 4(y-1)}{y(y-1)} = \dfrac{3y - 4y + 4}{y(y-1)} = \dfrac{4-y}{y(y-1)}$

It is extremely important that you read through this working very carefully and understand it. You must practise some similar examples from your textbook.

In part (b), be careful when simplifying the numerator $3y - 4(y-1)$, leading to $3y - 4y + 4$. Remember that $-4 \times -1 = +4$. Many students often make a mistake when multiplying a bracket by a negative number. It would be correct to write the final numerator in part (b) as $-y + 4$ rather than $4 - y$.

These questions are worth three marks each, one for the correct denominator, one for the correct numerators, and one for simplifying the numerator.

Example Express $\dfrac{2}{a} - \dfrac{3}{a^2}$ $(a \neq 0)$ as a fraction in its simplest form.

(Solution) Be careful here. The LCM is a^2 because a is a factor of a^2.

Hence $\dfrac{2}{a} - \dfrac{3}{a^2} = \dfrac{2a}{a^2} - \dfrac{3}{a^2} = \dfrac{2a-3}{a^2}$.

Multiplication of Fractions

To multiply fractions together, you should multiply the numerators together, multiply the denominators together and then simplify the fraction, if possible, by cancelling.

Example Express $\dfrac{2s^3}{t^2} \times \dfrac{t}{4s^2}$ as a fraction in its simplest form.

(Solution) $\dfrac{2s^3}{t^2} \times \dfrac{t}{4s^2} = \dfrac{2s^3 \times t}{t^2 \times 4s^2} = \dfrac{2 \times s \times s \times s \times t}{t \times t \times 4 \times s \times s} = \dfrac{\cancel{2}^1 \times \cancel{s} \times \cancel{s} \times s \times \cancel{t}}{\cancel{t} \times t \times \cancel{4}^2 \times \cancel{s} \times \cancel{s}} = \dfrac{s}{2t}$.

This is worth two marks, one for the first step, and one for going on to the correct answer.

Division of Fractions

To divide two fractions, we leave the first fraction as it is, change the division sign to a multiplication sign, turn the second fraction upside down, then do as a multiplication.

Example Express $\dfrac{a}{b} \div \dfrac{a^2}{3b}$ as a fraction in its simplest form.

(Solution) $\dfrac{a}{b} \div \dfrac{a^2}{3b} = \dfrac{a}{b} \times \dfrac{3b}{a^2} = \dfrac{a \times 3b}{b \times a^2} = \dfrac{a \times 3 \times b}{b \times a \times a} = \dfrac{\cancel{a} \times 3 \times \cancel{b}}{\cancel{b} \times \cancel{a} \times a} = \dfrac{3}{a}$.

This is worth three marks. The extra mark is for the first step.

Practise some multiplication and division examples. Students who have not prepared properly for the exam often get the four processes mixed up. Be clear about how to tackle each process before your exam. Here are some examples for you to try.

For Practice Express each of the following as a fraction in its simplest form.

a) $\dfrac{3}{x} + \dfrac{2}{x+4}$ $(x \neq 0, x \neq -4)$ b) $\dfrac{x^2}{y} \times \dfrac{y^2}{2x}$

c) $\dfrac{1}{x-2} - \dfrac{2}{x+1}$ $(x \neq 2, x \neq -1)$ d) $\dfrac{2h}{5} \div \dfrac{h^2}{15}$.

Answers: a) $\dfrac{5x+12}{x(x+4)}$ b) $\dfrac{xy}{2}$ c) $\dfrac{5-x}{(x-2)(x+1)}$ d) $\dfrac{6}{h}$.

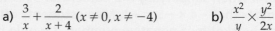

Well done if you got them all right! Check how to correct any errors.

Changing the Subject of a Formula

There are many formulae in maths. One example is the formula for the volume of a cylinder, $V = \pi r^2 h$, which you met in Unit 1. V is the subject of this formula. Suppose, however, you knew the volume but did not know the radius. Can you change the subject of the formula to r? If you can your progress is excellent!

In fact $V = \pi r^2 h \Rightarrow \pi r^2 h = V \Rightarrow r^2 = \dfrac{V}{\pi h} \Rightarrow r = \sqrt{\dfrac{V}{\pi h}}$.

In this section we shall look at different examples of changing the subject of a formula. There are several ways of teaching this, so use the method you are comfortable with. Some students work backwards from the original formula, undoing each step as they go. I prefer to think of changing the subject of a formula as solving an equation with letters in it instead of numbers. You can use the same methods you use to solve ordinary equations. Study the examples below, making sure you follow each step, and then practise similar examples from your textbook.

Example

a) Change the subject of the formula $w = \dfrac{xy}{z}$ to y.

b) Change the subject of the formula $A = b^2 c - d$ to h

c) Change the subject of the formula $c = \dfrac{4(d + e)}{f}$ to d.

Part (a) is worth two marks, parts (b) and (c) are worth three marks each.

(Solution) a) $w = \dfrac{xy}{z} \Rightarrow \dfrac{xy}{z} = w$ (swap the sides round)

$\Rightarrow xy = wz$ (cross-multiply)

$\Rightarrow y = \dfrac{wz}{x}$ (divide)

b) $A = b^2 c - d \Rightarrow b^2 c - d = A$ (swap the sides round)

$\Rightarrow b^2 c = A + d$ (add d to both sides)

$\Rightarrow b^2 = \dfrac{A + d}{c}$ (divide both sides by c)

$\Rightarrow b = \sqrt{\dfrac{A + d}{c}}$ (take the square root of both sides)

c) $c = \dfrac{4(d + e)}{f} \Rightarrow \dfrac{4(d + e)}{f} = c$ (swap the sides round)

$\Rightarrow 4(d + e) = cf$ (cross-multiply)

$\Rightarrow d + e = \dfrac{cf}{4}$ (divide both sides by 4)

$\Rightarrow d = \dfrac{cf}{4} - e$ (subtract e from both sides)

In part (c), there is an alternative method after you cross-multiply. You could remove the brackets: $4(d + e) = cf \Rightarrow 4d + 4e = cf \Rightarrow 4d = cf - 4e$. This leads to $d = \dfrac{cf - 4e}{4}$. This answer is equivalent to the earlier answer. Both would receive full marks.

Part (b) above is a common type of exam question. Remember at the end to 'cover' the entire fraction with the square root symbol.

Finally we look at two more examples involving fractions.

Example

a) Change the subject of the formula $m = \dfrac{1}{2}(a + b)$ to a.

b) Change the subject of the formula $s = ut + \dfrac{1}{2}at^2$ to a.

(Solution) a) $m = \dfrac{1}{2}(a + b) \Rightarrow 2m = a + b$ (cross-multiply)

$\Rightarrow a = 2m - b$ (subtract b from both sides)

This is worth two marks.

b) $s = ut + \dfrac{1}{2}at^2 \Rightarrow ut + \dfrac{1}{2}at^2 = s$ (swap sides round)

$\Rightarrow \dfrac{1}{2}at^2 = s - ut$ (subtract ut from both sides)

$\Rightarrow at^2 = 2(s - ut)$ (cross-multiply)

$\Rightarrow a = \dfrac{2(s - ut)}{t^2}$ (divide both sides by t^2)

This is worth three marks.

In part (b), you cannot cross-multiply immediately because of the $+$ sign in the equation. Examples (c) (earlier) and (b) (here) would be considered very difficult. If you can follow these parts, then you should have no problems in the exam!

Surds

You should be aware that surds are square roots which do not have an exact value, e.g. $\sqrt{7}$. Note that $\sqrt{9}$ is not a surd because $\sqrt{9} = 3$.

Questions on surds will usually appear in Paper 1. Never use a calculator for them.

Simplifying Surds

To simplify a surd you must find a factor which is a **perfect square**. The perfect squares to look out for are 4, 9, 16, 25, 36, 49, 64, 81, 100. If there is more than one perfect square, choose the largest.

Example

Simplify   a) $\sqrt{50}$ b) $\sqrt{72}$.

(Solution)   a) $\sqrt{50} = \sqrt{25 \times 2} = 5\sqrt{2}$.   b) $\sqrt{72} = \sqrt{36 \times 2} = 6\sqrt{2}$.

Careful! In part (b), some students would write $\sqrt{72} = \sqrt{9 \times 8} = 3\sqrt{8}$. This is true, but not fully simplified because $3\sqrt{8} = 3 \times \sqrt{4 \times 2} = 3 \times 2\sqrt{2} = 6\sqrt{2}$. So it is important that you find the largest perfect square factor when simplifying a surd.

 A common type of exam question requires simplifying an expression involving the addition and subtraction of surds. To do this, you must first be able to simplify surds.

Example

a) Express $\sqrt{27} + \sqrt{12} - 4\sqrt{3}$ as a surd in its simplest form.

b) Express $\sqrt{8} - \sqrt{2} + \sqrt{18}$ as a surd in its simplest form.

(Solution) a) $\sqrt{27} + \sqrt{12} - 4\sqrt{3} = \sqrt{9 \times 3} + \sqrt{4 \times 3} - 4\sqrt{3} = 3\sqrt{3} + 2\sqrt{3} - 4\sqrt{3}$
$$= (3 + 2 - 4)\sqrt{3}$$
$$= \sqrt{3}.$$

b) $\sqrt{8} - \sqrt{2} + \sqrt{18} = \sqrt{4 \times 2} - \sqrt{2} + \sqrt{9 \times 2} = 2\sqrt{2} - \sqrt{2} + 3\sqrt{2}$
$$= (2 - 1 + 3)\sqrt{2}$$
$$= 4\sqrt{2}.$$

 Both examples are worth three marks, one for each surd you have to simplify, and one for collecting like surds at the end of the working. In both parts, there is a surd which you do not need to simplify, $\sqrt{3}$ in part (a) and $\sqrt{2}$ in part (b). These usually give a hint about simplifying the other two surds in each part.

Multiplying and Dividing Surds

Example

a) Simplify $2\sqrt{3} \times 4\sqrt{15}$.

b) Simplify $\dfrac{\sqrt{24}}{\sqrt{2}}$.

c) If $\sqrt{a} = 4\sqrt{5}$, find the value of a.

(Solution) a) Multiply whole numbers and surds separately and then simplify the surd, if possible.
Hence $2\sqrt{3} \times 4\sqrt{15} = 2 \times 4 \times \sqrt{3} \times \sqrt{15} = 8 \times \sqrt{45} = 8 \times \sqrt{9 \times 5} = 8 \times 3\sqrt{5}$
$$= 24\sqrt{5}.$$

b) $\dfrac{\sqrt{24}}{\sqrt{2}} = \sqrt{\dfrac{24}{2}} = \sqrt{12} = \sqrt{4 \times 3} = 2\sqrt{3}.$

c) $\sqrt{a} = 4\sqrt{5} = \sqrt{16} \times \sqrt{5} = \sqrt{16 \times 5} = \sqrt{80} \Rightarrow a = 80.$

Rationalizing the Denominator

When a fraction has a surd in the denominator, it is common to be asked to express the fraction with a **rational denominator**. A question on this in the exam is worth two marks, one for the correct method and one for the correct answer.

It depends on the fact that when you multiply a surd by itself, you get a rational number, e.g. $\sqrt{2} \times \sqrt{2} = \sqrt{4} = 2$, $\sqrt{3} \times \sqrt{3} = 3$, etc.

Study the examples below and learn the method.

Example

a) Express $\dfrac{5}{\sqrt{2}}$ as a fraction with a rational denominator.

b) Express $\dfrac{6}{\sqrt{3}}$ with a rational denominator.

(Solution) a) $\dfrac{5}{\sqrt{2}} = \dfrac{5}{\sqrt{2}} \times \dfrac{\sqrt{2}}{\sqrt{2}} = \dfrac{5\sqrt{2}}{2}$
b) $\dfrac{6}{\sqrt{3}} = \dfrac{6}{\sqrt{3}} \times \dfrac{\sqrt{3}}{\sqrt{3}} = \dfrac{6\sqrt{3}}{3} = 2\sqrt{3}.$

Indices

This is the final topic in this outcome. Indices (plural of index) are powers. In the number 3^5, '5' is the index and '3' is called the base. Remember that 3^5 means $3 \times 3 \times 3 \times 3 \times 3 = 243$. It is vital that you know the **Laws of Indices**.

Remember

Law 1. $a^p \times a^q = a^{p+q}$. When **multiplying** numbers raised to powers, **add** indices.

Examples. a) $x^5 \times x^4 = x^{5+4} = x^9$

b) $a^{\frac{7}{3}} \times a^{\frac{2}{3}} = a^{\frac{7}{3}+\frac{2}{3}} = a^{\frac{9}{3}} = a^3.$

Make sure that you follow the addition of fractions in (b).

Law 2. $a^p \div a^q = a^{p-q}$. When **dividing** numbers raised to powers, **subtract** indices.

Example. $c^{\frac{9}{2}} \div c^{\frac{5}{2}} = c^{\frac{9}{2}-\frac{5}{2}} = c^{\frac{4}{2}} = c^2.$

Carefully check the subtraction of fractions.

Law 3. $(a^p)^q = a^{pq}$. When **raising** a number to a power to another power, **multiply** indices.

Example. $(t^5)^3 = t^{5 \times 3} = t^{15}.$

Law 4. $(ab)^n = a^n b^n.$

Example. $(k^2 l)^3 = (k^2)^3 l^3 = k^6 l^3.$

Notice how Law 3 was used to complete the answer.

Remember *continued* ⟩

Remember

Law 5. $a^{-n} = \dfrac{1}{a^n}$.

This helps us to deal with **negative** indices.

Example   Express $u^{-5} \times u^2$ with a positive index.

(Solution) $u^{-5} \times u^2 = u^{-5+2} = u^{-3} = \dfrac{1}{u^3}$.

Law 6. $a^0 = 1$. Any number to the power **zero** is equal to 1.

Example. $h^{\frac{1}{2}} \times h^{-\frac{1}{2}} = h^{\frac{1}{2} + (-\frac{1}{2})} = h^0 = 1$.

Law 7. $a^{\frac{p}{q}} = \sqrt[q]{a^p}$. This helps us to deal with **fractional** indices.

Example   a) Evaluate $8^{\frac{2}{3}}$.   b) Evaluate $9^{\frac{3}{2}}$.

(Solution)   **a)** $8^{\frac{2}{3}} = \sqrt[3]{8^2} = (\sqrt[3]{8})^2 = 2^2 = 4$.

            b) $9^{\frac{3}{2}} = \sqrt[2]{9^3} = (\sqrt{9})^3 = 3^3 = 27$.

When evaluating fractional indices, always calculate the root first, then the power. It is easier in that order. $\sqrt[3]{8} = 2$ because $2^3 = 8$. The symbol $\sqrt[2]{\ }$ is the same as $\sqrt{\ }$ and means 'square root'.

Here are some other roots which could prove useful:

$\sqrt[3]{27} = 3$, $\sqrt[3]{64} = 4$, $\sqrt[4]{16} = 2$.

Here is a **summary** of the Laws of Indices.

Summary

	Law	Illustration
1	$a^p \times a^q = a^{p+q}$	$x^5 \times x^2 = x^{5+2} = x^7$
2	$a^p \div a^q = a^{p-q}$	$y^{10} \div y^3 = y^{10-3} = y^7$
3	$(a^p)^q = a^{pq}$	$(z^3)^5 = z^{3\times5} = z^{15}$
4	$(ab)^n = a^n b^n$	$(x^2 y^3)^4 = (x^2)^4 (y^3)^4 = x^8 y^{12}$
5	$a^{-n} = \dfrac{1}{a^n}$	$m^{-2} = \dfrac{1}{m^2}$
6	$a^0 = 1$	$7^0 = 1$
7	$a^{\frac{p}{q}} = \sqrt[q]{a^p}$	$f^{\frac{2}{3}} = \sqrt[3]{f^2}$

! Make sure you know all of these Laws of Indices. They will help you to do all exam questions on indices. We shall look at some typical exam questions next.

Example

a) Simplify $(h^3)^4 \times h^5$

b) Simplify $10w^{\frac{7}{2}} \div 2w^{\frac{1}{2}}$

c) Simplify $\dfrac{b^{\frac{5}{2}} \times b^{-\frac{1}{2}}}{b}$

d) Simplify $5e^3 \times 4e^2$

(Solution)

a) $(h^3)^4 \times h^5 = h^{3\times4} \times h^5 = h^{12} \times h^5 = h^{12+5} = h^{17}$.

b) $10w^{\frac{7}{2}} \div 2w^{\frac{1}{2}} = 5w^{\frac{7}{2}-\frac{1}{2}} = 5w^{\frac{6}{2}} = 5w^3$ (Note: divide 10 by 2 (=5)).

c) $\dfrac{b^{\frac{5}{2}} \times b^{-\frac{1}{2}}}{b} = \dfrac{b^{\frac{5}{2}+(-\frac{1}{2})}}{b} = \dfrac{b^{\frac{4}{2}}}{b} = \dfrac{b^2}{b} = b^{2-1} = b^1 = b$.

d) $5e^3 \times 4e^2 = 20e^{3+2} = 20e^5$ (Note: multiply 5 by 4 (=20)).

! As you work through the above examples, refer to the Laws of Indices to help your understanding. Be careful when adding and subtracting fractions, e.g. $\dfrac{7}{2} - \dfrac{1}{2} = \dfrac{6}{2} = 3$. Note also in part (c) that $b^1 = b$.

Example

a) Express $a^{\frac{2}{3}}(a^{\frac{1}{3}} - a^{-\frac{2}{3}})$ in its simplest form.

b) Express $16^{-\frac{3}{4}}$ as a fraction in its simplest form.

(Solution) a) $a^{\frac{2}{3}}(a^{\frac{1}{3}} - a^{-\frac{2}{3}}) = a^{\frac{2}{3}} \times a^{\frac{1}{3}} - a^{\frac{2}{3}} \times a^{-\frac{2}{3}} = a^{\frac{2}{3}+\frac{1}{3}} - a^{\frac{2}{3}+(-\frac{2}{3})} = a^{\frac{3}{3}} - a^0 = a - 1$.

! There is a lot to take in here, although it is worth only two marks. Multiply out brackets as normal, and use the first Law of Indices. Remember that $a^1 = a$ and that $a^0 = 1$.

b) $16^{-\frac{3}{4}} = \dfrac{1}{16^{\frac{3}{4}}} = \dfrac{1}{\sqrt[4]{16^3}} = \dfrac{1}{2^3} = \dfrac{1}{8}$.

(This part uses the 5th and 7th laws. When evaluating $16^{\frac{3}{4}}$, think of $\sqrt[4]{16^3}$ as being $(\sqrt[4]{16})^3$. Then $\sqrt[4]{16} = 2$, because $2^4 = 16$, and $2^3 = 8$.)

Hints and Tips

Be patient when studying this algebra outcome. It takes time and perseverance to succeed at it.

Chapter 13

QUADRATIC FUNCTIONS

In Unit 1, we studied equations such as $y = 2x + 3$. These equations led to straight line graphs. In this outcome we shall study quadratic equations such as $y = 2x^2$, $y = (x - 2)^2 + 5$, and $y = x^2 + 5x + 7$. Equations of this type, all of which contain an x^2 term, lead to graphs called *parabolas*. A parabola is a symmetrical graph with either a maximum or a minimum turning point.

Here we shall concentrate on typical exam questions and leave most of the theory, important though it is, to your textbook.

There are usually between seven and eleven marks on parabolas and quadratic equations in the exam.

Equations of the Form $y = kx^2$

The equation $y = kx^2$ is the equation of a parabola with a turning point at the Origin.

Example

The diagram below shows the graph of $y = kx^2$.

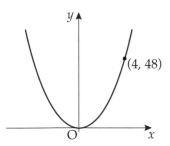

Find the value of k.

(Solution)   Substitute (4, 48) into the equation $y = kx^2$, 4 for x and 48 for y.

Hence $y = kx^2 \Rightarrow 48 = k \times 4^2 \Rightarrow 48 = k \times 16$

$$\Rightarrow k = \frac{48}{16}$$

$$\Rightarrow k = 3.$$

This type of question is quite common. It is worth two marks.

Equations of the Form $y = (x + a)^2 + b$

The expression $(x + a)^2 + b$ has a minimum value of b when $x = -a$.

Example

Find the minimum value of $(x + 3)^2 + 7$ and write down the coordinates of the turning point of the parabola with equation $y = (x + 3)^2 + 7$

(Solution) $(x + 3)^2 + 7$ has a minimum value of 7 when $x = -3$.

Why? Because when $x = -3$, $(x + 3)^2 + 7$ becomes $(-3 + 3)^2 + 7 = 0 + 7 = 7$.

If x was replaced with any value other than -3, $(x + 3)^2$ would be greater than zero.

The coordinates of the **minimum** turning point of the parabola with equation $y = (x + 3)^2 + 7$ are therefore $(-3, 7)$.

Equations of the Form $y = k(x + a)^2 + b$

This equation looks difficult, but exam questions based on it are usually straightforward. In the exam, the only values of k which can occur are ± 1. As a result you will meet equations of the type $y = (x + 2)^2 + 3$, (where $k = 1$), or $y = 8 - (x - 2)^2$, (where $k = -1$).

Example

The diagram below shows part of the graph of $y = (x - 4)^2 - 9$.

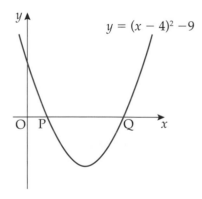

a) State the coordinates of the minimum turning point.
b) State the equation of the axis of symmetry.
c) P has coordinates $(1, 0)$. What are the coordinates of Q?

Example *continued* ➤

Example *continued*

In part (a), $(x - 4)^2 - 9$ has a minimum value when $x = 4$. Then $y = (4 - 4)^2 - 9 = -9$. The value of x, which makes the bracket equal to zero, will give you the x-coordinate of the turning point. The remaining term, -9, gives you the y-coordinate of the turning point. Therefore $(4, -9)$ is the turning point. Always check with the diagram to see if your answer is sensible. You can see from the diagram that the turning point is in the 4th quadrant so $(4, -9)$ is sensible.

In part (b), the axis of symmetry is a vertical line through the turning point. Its equation is $x = 4$. This answer must be consistent with the x-coordinate of the turning point. Don't forget to put '$x =$' before the number.

Part (c) is done by symmetry. From $x = 1$ at P to $x = 4$ is 3 units, so there will be another 3 units from $x = 4$ to point Q, leading to $(7, 0)$.

(Solution)   **a)** $(4, -9)$ **b)** $x = 4$ **c)** Q is $(7, 0)$.

Example

The diagram below shows part of the graph of $y = 13 - (x + 2)^2$.

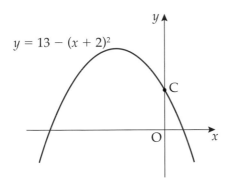

$y = 13 - (x + 2)^2$

a) Write down the coordinates of the maximum turning point.
b) State the equation of the axis of symmetry.
c) Find the coordinates of C, the point where the graph cuts the y-axis.

(Solution)   **a)** $(-2, 13)$.
            b) $x = -2$.
            c) Graph cuts y-axis where $x = 0 \Rightarrow y = 13 - (0 + 2)^2$
$$\Rightarrow y = 13 - 4$$
$$\Rightarrow y = 9$$

Hence C is $(0, 9)$.

With a little practice, you should be able to cope well with examples like these. It will be worth your while to practise as many as possible, as they appear *often* in the exam.

Quadratic Equations

Quadratic equations can be solved by

    a) graphical methods;
    b) factorisation;
    c) the quadratic formula;
    d) using a graphical calculator.

Drawing an accurate graph of a quadratic function can be time-consuming and is unlikely to be expected in an exam situation.

You are allowed to use a graphical calculator in the exam, although very few students do. Even if you are confident with the graphical calculator, I would recommend that you use methods (b) or (c) above to solve quadratic equations. A graphical calculator could then be used to check your answers. However, if you are determined to use one, you must show the graph you have drawn, with the equation and the axes labelled, and also indicate the position of the roots, before giving the solutions.

We shall concentrate here on methods (b) and (c).

Solving Quadratic Equations by Factorisation

Example

Solve the equation $x^2 - 4x - 45 = 0$.

(Solution) $x^2 - 4x - 45 = 0 \Rightarrow (x + 5)(x - 9) = 0$
$$\Rightarrow \text{ either } x + 5 = 0 \text{ or } x - 9 = 0$$
$$\Rightarrow x = -5 \text{ or } x = 9.$$

Key Points

Obviously, to do this, you must be able to factorise a trinomial. The two solutions $x = -5$ or $x = 9$ are called the **roots** of the equation. These roots tell us where the parabola with equation $y = x^2 - 4x - 45$ cuts the x-axis. If you substitute $x = -5$ or $x = 9$ into this equation, then $y = 0$. Therefore this parabola would cut the x-axis at $(-5, 0)$ and $(9, 0)$.

Example

Find the roots the equation $x^2 - 3x = 0$.

(Solution)   Using a common factor, $x^2 - 3x = 0 \Rightarrow x(x - 3) = 0$
$$\Rightarrow \text{ either } x = 0 \text{ or } x - 3 = 0$$
$$\Rightarrow x = 0 \text{ or } x = 3.$$

Remember that the *type* of factorisation could be a common factor, a difference of two squares or a trinomial. Now we shall solve a more difficult problem using factorisation.

Example

The diagram shows a square room in a glasshouse, which consists of a rectangular soft play area with a path on three sides.
The room measures 5 metres by 5 metres.
The width of the path is x metres.

a) State the length and breadth of the soft play area.

b) Show that the area, A square metres, of the soft play area is given by
$$A = 2x^2 - 15x + 25$$

c) The area of the soft play area is 12 square metres. By using the equation in part (b), find the width of the path.

(Solution) a) Length $= 5 - x$, Breadth $= 5 - 2x$.   (or vice versa)

 b) Area $=$ Length $\times$ Breadth $= (5 - x)(5 - 2x)$
$$= 5(5 - 2x) - x(5 - 2x)$$
$$= 25 - 10x - 5x + 2x^2$$
$$= 25 - 15x + 2x^2$$

So $A = 2x^2 - 15x + 25$.

c) If Area $= 12$, $2x^2 - 15x + 25 = 12$
$$\Rightarrow 2x^2 - 15x + 25 - 12 = 0$$
$$\Rightarrow 2x^2 - 15x + 13 = 0$$
$$\Rightarrow (2x - 13)(x - 1) = 0$$
$$\Rightarrow \text{either } 2x - 13 = 0 \text{ or } x - 1 = 0$$
$$\Rightarrow x = \frac{13}{2} \text{ or } x = 1.$$

Hence width $= 1$ metre (since $\frac{13}{2} = 6{\cdot}5$ metres, which is wider than the room).

 Parts (a) and (b) are actually Unit 1 work. Only part (c) is from Unit 3. However, you occasionally get questions which have parts from different Units. In part (c), you **must** make the right hand side of the equation equal to zero before you factorise. Notice also how one of the two solutions was not possible. This often happens in real-life problems too. For example, you might sometimes get a negative answer for a length, so watch out.

Solving Quadratic Equations by the Quadratic Formula

Remember

The quadratic equation $ax^2 + bx + c = 0$ can be solved using the quadratic formula
$$x = \frac{-b \pm \sqrt{(b^2 - 4ac)}}{2a}.$$

This formula is normally used to solve quadratic equations which cannot be solved by factorisation, although it can be used to solve *any* quadratic equation.

There has been a question in the exam on the quadratic formula most years. If you are asked to solve an equation, **giving the roots correct to one decimal place**, then you should use the quadratic formula. You will be able to use a calculator and the question will be worth four marks. As usual, when copying a formula from the formula sheet, take care. In particular, make sure that the line separating the numerator and the denominator of the fraction goes all the way across throughout your calculation.

Example

Solve the equation $2x^2 - 5x - 4 = 0$ giving the roots correct to one decimal place.

(Solution) $a = 2, b = -5, c = -4$

$$x = \frac{-b \pm \sqrt{(b^2 - 4ac)}}{2a} \Rightarrow x = \frac{-(-5) \pm \sqrt{((-5)^2 - 4 \times 2 \times (-4))}}{2 \times 2}$$

$$\Rightarrow x = \frac{5 \pm \sqrt{(25 + 32)}}{4}$$

$$\Rightarrow x = \frac{5 \pm \sqrt{57}}{4}$$

$$\Rightarrow x = \frac{5 + \sqrt{57}}{4} \ or \ \frac{5 - \sqrt{57}}{4}$$

$$\Rightarrow x = \frac{12 \cdot 55}{4} \ or \ \frac{-2 \cdot 55}{4}$$

$$\Rightarrow x = 3 \cdot 1 \ or \ -0 \cdot 6 \ \text{(correct to 1 d.p.)}$$

Hints and Tips

There is 1 mark for calculating $b^2 - 4ac$ correctly (57 in this case). If you get a negative answer for $b^2 - 4ac$, you have probably made a mistake, so check your working, remembering that b^2 must be positive.

Some students forget to round the answers to one decimal place (one number after the decimal point) in their answers, and lose a mark. Don't be one of them. There is information about rounding to decimal places in Chapter 2.

When doing calculations such as $\frac{5 + \sqrt{57}}{4}$, always write down the answer to the numerator **before** dividing, as shown above. Because if you key in $5 + \sqrt{57} \div 4$ into your calculator, it will only divide $\sqrt{57}$ by 4, not $5 + \sqrt{57}$ by 4.

It is important that you are confident about using this formula correctly, so make sure you practise a variety of examples from your textbook.

Finally in this section, we return to the parabola problem, with a more difficult example!

Example

a) Factorise $x^2 + 2x - 15$.
b) Hence write down the roots of the equation
$$x^2 + 2x - 15 = 0.$$
c) The graph of $y = x^2 + 2x - 15$ is shown in the diagram.

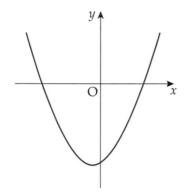

Find the coordinates of the turning point.

(Solution) a) $x^2 + 2x - 15 = (x + 5)(x - 3)$.

b) $x^2 + 2x - 15 = 0 \Rightarrow (x + 5)(x - 3) = 0$
$$\Rightarrow x + 5 = 0 \ \text{or} \ x - 3 = 0$$
$$\Rightarrow x = -5 \ \text{or} \ x = 3$$
$$\Rightarrow \text{roots are } -5 \text{ or } 3.$$

c) The axis of symmetry is midway between -5 and 3, i.e. at $x = -1$.
Substitute $x = -1$ into $y = x^2 + 2x - 15$
$$\Rightarrow y = (-1)^2 + 2 \times (-1) - 15$$
$$\Rightarrow y = 1 - 2 - 15$$
$$\Rightarrow y = -16.$$
Hence coordinates of the turning point are $(-1, -16)$.

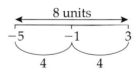

A problem of this difficulty could be worth up to six marks.

In part (b), you could simply write down the roots, -5 and 3, without working.

For this outcome, the most important thing is to study the examples and be able to attempt similar questions in the exam, even if you are a bit unsure about all of the theory behind them.

FURTHER TRIGONOMETRY

In this, the final outcome in Unit 3, we shall study trigonometric graphs, equations and identities. Calculators are allowed when solving equations. They are not required for graphs or identities. There are usually between five and seven marks in total on this outcome.

The Graphs of $\sin x°$, $\cos x°$ and $\tan x°$

You must be able to recognise the graphs of $y = \sin x°$, $y = \cos x°$, and $y = \tan x°$.

You should be able to sketch them for the interval $0 \leqslant x \leqslant 360$ and you should know that the maximum and minimum values of both $y = \sin x°$ and $y = \cos x°$ are 1 and -1 respectively.

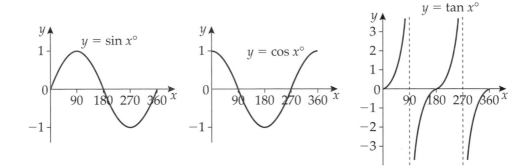

Related Graphs

You must also be able to **sketch** and **identify** trigonometric graphs involving
a) amplitude and multiple angle
b) phase angle.

We shall look at amplitude and multiple angle first, with reference to the graphs of $y = a \sin bx°$ and $y = a \cos bx°$.

In these two equations, the letter **before** sin and cos, a, is the *amplitude* of the graph and changes the maximum and minimum values from 1 and -1 to a and $-a$.

The letter **after** sin and cos, b, tells you how many cycles of the graph there will be between $0°$ and $360°$.

Example

The graph of $y = a \cos bx°$ is shown in the diagram.
State the values of a and b.

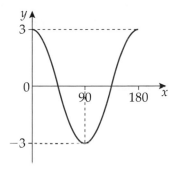

(Solution) $a = 3, b = 2$.

 This example is worth two marks. Note that $b = 2$ because there would be two full cycles of the cosine graph between 0° and 360°. This is a typical question from Paper 1 and is easy if you have prepared properly.

You could be asked to sketch these more complex graphs. But before you can do this, you **must** be able to sketch the graphs of $y = \sin x°$ and $y = \cos x°$.

Example

Sketch the graph of

$$y = 4 \sin 2x°, 0 \leqslant x \leqslant 360.$$

(Solution)

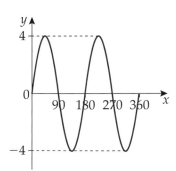

 Make sure you begin by drawing a sine graph. It starts at the origin. Mark clearly its maximum and minimum values at 4 and −4. Then draw 2 cycles of the sine graph between 0° and 360°. Do not draw one cycle between 0° and 180° and stop there. Try and draw a *smooth* curve, showing where it crosses the *x*-axis. It would be a good idea to practise drawing sine and cosine graphs in advance.

A question like this would be worth three marks.

Now we shall look at graphs involving a phase angle, i.e. the graphs of $y = \sin(x - a)°$ and $y = \cos(x - a)°$. **The term a is called the phase angle. It has the effect of moving the graphs of $y = \sin x°$ and $y = \cos x°$, $a°$ to the right.**

Example

The graph shown below has an equation in the form $y = \cos(x - a)°$.

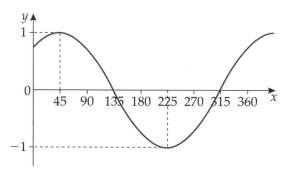

Write down the value of a.

(Solution) $a = 45$.

Here the cosine graph has been moved 45° to the right. This is worth one mark.

Periodicity

All the trigonometric graphs we have met have a repeating pattern. The **period** is the length of one cycle of the graph. Remember that the period of $y = \sin x°$ is 360°; the period of $y = \cos x°$ is 360°; but the period of $y = \tan x°$ is 180°.

The period will change if there is a multiple angle, e.g. for the graph of $y = a \sin bx°$, the period will be $(360 \div b)°$. Note that the amplitude a does not affect the period.

Example

State the period of $y = 3 \tan 4x°$.

(Solution)   The period of $y = \tan x°$ is 180°
$\Rightarrow$ the period of $y = 3 \tan 4x°$ is $(180 \div 4) = 45°$.

Now, we shall study the most demanding part of this outcome, trigonometric equations.

Trigonometric Equations

There has been a trigonometric equation in every Intermediate 2 exam so far, either a straightforward equation worth three marks or as part of a problem worth between four and six marks.

You should start by re-arranging the equation into the form $\sin x° = ...$, $\cos x° = ...$ or $\tan x° = ...$ Then you must use the information from Chapter 8, to find **two** solutions in the interval $0 \leqslant x \leqslant 360$.

Study the two examples below, and have your calculator ready (in degree mode).

All the equations you will meet will be similar to these.

Example Solve the equation

$$6 \sin x° - 5 = 0, \ 0 \leqslant x \leqslant 360.$$

(Solution) $6 \sin x° - 5 = 0$

$\Rightarrow 6 \sin x° = 5$

$\Rightarrow \sin x° = \dfrac{5}{6} = 0{\cdot}833$

Related angle: Now use $\sin^{-1} 0{\cdot}833 = 56{\cdot}4°$.

Now use the table to find the correct quadrants.

The **Sine** is **positive** in 1st, 2nd quadrants

Hence $x = 56{\cdot}4$ or $(180 - 56{\cdot}4)$

$\qquad = 56{\cdot}4$ or $123{\cdot}6$.

SIN	ALL
$180 - A$	A
TAN	COS
$180 + A$	$360 - A$

Example Solve the equation

$$5 \tan x° + 3 = 0, \ 0 \leqslant x \leqslant 360.$$

(Solution) $5 \tan x° + 3 = 0$

$\Rightarrow 5 \tan x° = -3$

$\Rightarrow \tan x° = -\dfrac{3}{5} = -0{\cdot}6.$

Related angle: Now use $\tan^{-1} 0{\cdot}6 = 31{\cdot}0°$.

(Note: omit the minus sign in front of $0{\cdot}6$ at this stage.)

Now use the table to find the correct quadrants.

The **Tangent** is **negative** in 2nd and 4th quadrants.

Hence $x = (180 - 31{\cdot}0)$ or $(360 - 31{\cdot}0)$

$\qquad = 149$ or 329.

SIN	ALL
$180 - A$	A
TAN	COS
$180 + A$	$360 - A$

Hints and Tips

Make sure you learn the **all, sin, tan, cos** table and the rules for each quadrant.

Remember that the related angle must be in the first quadrant. That is why the minus sign is 'omitted' in the second example.

With enough practice, you should have no problems with this type of equation.

It is possible, however, that a trigonometric equation could appear in a more general problem form.

Example

A 'speed trap' at S is tracking cars as they move along a straight road. It sweeps through an angle of $a°$. The distance of a car from a school gate, d metres, is calculated using the formula

$$d = 20 \tan a° + 100.$$

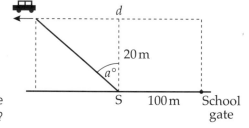

a) How far from the school gate is a car when the speed trap has swept through an angle of 40°?

b) Find two possible angles when the distance from the school gate is 140 metres.

c) Which one of these angles is actually possible? Give a reason for your answer.

(Solution) a) Substitute $a = 40$ into $d = 20 \tan a° + 100$
$$\Rightarrow d = 20 \tan 40° + 100$$
$$= 116 \cdot 8.$$
Hence the distance is 116·8 metres.

b) Substitute $d = 140$ into $d = 20 \tan a° + 100$
$$\Rightarrow 140 = 20 \tan a° + 100$$
$$\Rightarrow 20 \tan a° = 140 - 100$$
$$\Rightarrow 20 \tan a° = 40$$
$$\Rightarrow \tan a° = \frac{40}{20} = 2.$$

Related angle: Now use $\tan^{-1} 2 = 63 \cdot 4°$.
Now use the table to find the correct quadrants.
The **Tangent** is **positive** in 1st and 3rd quadrants.
Hence $x = 63 \cdot 4$ or $(180 + 63 \cdot 4)$
$$= 63 \cdot 4 \text{ or } 243 \cdot 4.$$

SIN	ALL
180 − A	A
TAN	COS
180 + A	360 − A

c) Because the road is straight, the angle must be less than 90°. Therefore $x = 63 \cdot 4$.

This example is difficult and would be worth seven marks (2 + 4 + 1). However, part (a) is basically substitution into a formula followed by a calculation which can be done in one go on your calculator. Part (b) is a trigonometric equation. Many students are put off by the wording and diagrams in questions like this. If you can see through all of the 'surrounding information' the question is often not as difficult as it first seems.

Identities

This is the final topic in this Unit. It is tested occasionally and usually always poorly attempted by students.

You should have met the formulae

$$\tan A = \frac{\sin A}{\cos A} \quad \text{and} \quad \sin^2 A + \cos^2 A = 1$$

These formulae are called trigonometric identities. This means that they are true for all values of A. These two identities can be used to prove other identities.

It is important to note that the second identity can be written as

$$\sin^2 A = 1 - \cos^2 A \quad \text{or} \quad \cos^2 A = 1 - \sin^2 A$$

Familiar with these formulae, you should be able to attempt some proofs!

Example

Show that:

$$\cos x° \tan x° = \sin x°.$$

Hints and Tips

Trigonometric questions of this type beginning with 'Show that' or 'Prove that' can be attempted by starting with the left hand side of the equation, substituting in an appropriate formula (from the ones above) and then simplifying the expression to give (hopefully) the correct right hand side. Look at the solution below.

(Solution)   left hand side $= \cos x° \tan x°$

$$= \cos x° \frac{\sin x°}{\cos x°}$$

$$= \sin x°$$

$$= \text{right hand side.}$$

FURTHER TRIGONOMETRY

Example

Show that:

$$\cos^3 x° + \cos x° \sin^2 x° = \cos x°.$$

This appears extremely difficult. The only formula you know on the left hand side is $\sin^2 x° = 1 - \cos^2 x°$. Substitute it into the left hand side, however, and continue from there.

(Solution)   left hand side $= \cos^3 x° + \cos x° \sin^2 x°$
$= \cos^3 x° + \cos x° (1 - \cos^2 x°)$
$= \cos^3 x° + \cos x° - \cos^3 x°$
$= \cos x°$
$=$ right hand side.

The secret to proving identities is to know the four formulae given at the start of this section and to substitute any one of them into the left hand side.

This type of question could be presented in a slightly different way.

Example

Simplify:

$$\frac{\sin^2 B}{1 - \sin^2 B}.$$

(Solution)   Substitute $\cos^2 B$ for $1 - \sin^2 B$.

Hence $\dfrac{\sin^2 B}{1 - \sin^2 B} = \dfrac{\sin^2 B}{\cos^2 B} = \left(\dfrac{\sin B}{\cos B}\right)^2 = \tan^2 B.$

Learn the identities at the start of this section and practice a few examples from your textbook until your confidence improves. Once you get a few right answers, it won't seem so difficult.

This brings us to the end of Unit 3 **and** the end of the course. Hopefully you have learned enough to pass Intermediate 2, and with a high grade.

You now have to practise exam papers.

Miss out Chapters 15 to 18. They are for those students doing the Applications of Mathematics Unit.

Go straight to Appendix 1 now where you will find a typical Paper 1 to practise.

Chapter 15

CALCULATIONS WITH MONEY

 Do not read Chapters 15 to 18 if you are studying Unit 3. Go straight to Appendix 1.

In this outcome, important aspects of working life concerning money are studied: earning money, borrowing money, spending money and paying tax. Although most questions will appear in Paper 2, there could be some questions in Paper 1.

There will probably be between eight and eleven marks in total on this outcome.

Percentages

You **must** be able to calculate percentages, with and without a calculator. If, for example, you had to calculate 15% of £800 in Paper 1, could you do it?

The easiest way is to find 10% of £800 = £80; then 5% of £800 = £40; leading to the answer, 15% of £800 = £120.

(With a calculator, 15% of £800 = 0.15×800 = £120.)

We now look at some examples of topics related to earning money.

 ## *Overtime*

Overtime is a way of earning more than a basic wage, usually by working extra hours in the evening or at the weekend. The most common rates of overtime are *double time* ($2 \times$ basic wage) and *time and a half* ($1.5 \times$ basic wage).

To work out the *time and a half* rate, multiply by 1.5 if you are permitted a calculator. For example, if the basic rate is £4.80 per hour, then the time and a half rate is 1.5×4.80 = £7.20.

If you are asked this in the Non-calculator paper, halve £4.80 to get £2.40, and then add £4.80 and £2.40 for £7.20.

Example (No calculator allowed)

Davinder works part-time in a local shop.
He is paid at a rate of £5.20 per hour for weekdays and at time and a half for weekends.
During one week he works from 8 am till 1 pm **every** day except Monday.
Calculate his gross pay for that week.

(Solution)   8 am till 1 pm = 5 hours per day
Weekday Pay (Tue–Fri) = $(4 \times 5) \times 5.20 = 20 \times 5.20 = 2 \times 52$ = £104
(Overtime rate (time and a half) = $5.20 + 2.60$ = £7.80 per hour)
Overtime Pay (Sat–Sun) = $(2 \times 5) \times 7.80 = 10 \times 7.80$ = £78
Hence gross Pay for the week = £104 + £78 = £182.

Check all the working carefully, and make sure you understand it. This example is worth three marks (one for knowing how to calculate weekday pay, one for knowing how to calculate overtime pay, and one for doing all the calculations correctly).

Commission

Commission is a way of paying people a percentage of the amount of goods they sell. It encourages salespersons to sell as much as possible.

Example (Calculator allowed)

Elizabeth sells cosmetics. She is paid a basic salary of £12 450 per annum plus 2·5% commission on all her sales. Calculate her salary in a year in which she sold £162 000 worth of cosmetics.

(Solution)   Basic salary = £12 450.
Commission = 2·5% of sales
$$= 2\text{·}5\% \text{ of } £162\,000 = \frac{2\text{·}5}{100} \times 162\,000 = 0\text{·}025 \times 162\,000$$
$$= £4050.$$
Hence salary for the year = Basic salary + Commission = £(12 450 + 4050)
$$= £16\,500.$$

Payslips

Example

Robert Sinkala works as a hairdresser. His April payslip, shown below, is only partly completed.

Name R. Sinkala	Employee No. 003	Tax Code 404L	Month April
Basic Salary £1200	Bonus £	Overtime Nil	Gross Salary £
Nat. Insurance £72·50	Income Tax £213·75	Pension £	Total Deductions £
			Net Salary £

a) Robert is paid a bonus of 10 pence for each haircut he does. Calculate his gross salary for April if he does 640 haircuts during the month.
b) 6% of Robert's gross monthly salary is paid into his pension fund. Calculate Robert's net salary for April.

Example continued ➤

Example *continued*

(Solution) **a)** Bonus = 640 × 10 p = 6400 p = £64.
Hence gross salary = £(1200 + 64) = £1264.

b) Pension = 6% of Gross salary = 6% of £1264 = 0·06 × 1264 = £75·84.
Hence total deductions = £(72·50 + 213·75 + 75·84) = £362·09.
Net salary = Gross salary − Total deductions = £(1264 − 362·09) = £901·91.

This question is worth five marks, two for part (a) and three for part (b).

Income Tax

Questions on Income Tax are quite common. They are usually worth four marks. You must be prepared for them but you may use a calculator.

If you are asked to find the amount of income tax paid by someone, you must start by finding the **taxable income**.

Remember

Taxable Income = Income − Allowances

You do not pay tax on allowances.

Look at this example, and follow each step carefully.

Example

Dorothy earns £44 650 per year and has tax allowances totalling £6035. The rates of tax applicable are as follows.

TAXABLE INCOME (£)	RATE
On the first £34 800	20%
On any income over £34 800	40%

Calculate the amount of tax paid by Dorothy.

(Solution) Taxable income = Income − Allowances
= £(44 650 − 6035)
= £38 615.

Tax paid at the 20% rate = 20% of £34 800 = 0.2 × £34 800 = £6960.
Tax paid at the 40% rate = 40% of £(38 615 − 34 800)
= 40% of £3815
= 0·4 × £3815.
= £1526

Hence total tax paid = £(6960 + 1526) = £8486.

HOW TO PASS INTERMEDIATE 2 MATHS

Summary

If your taxable income is *less than £34 800*, you only pay 20% in tax.
If your taxable income is *more than £34 800*, you pay 20% (£6960) on the first £34 800, then 40% on £(taxable income − 34 800).

**Note that the rates of tax shown above were used in the year 2008–09.
Remember, too, that 1 year = 52 weeks, in case you are asked for weekly tax.**

Now we look at two topics related to Personal Spending.

Repayments

When money is borrowed, repayments have to be made, usually monthly.

The amount of the repayments depends on how much is borrowed, the annual percentage rate (APR) and the length of time over which any loan is taken.

Loan repayment tables give information about how much you will have to pay if you borrow money. These tables detail repayment amounts with or without payment protection.

The **cost of the loan** is the difference between the total repayments and the loan.

Questions on loan repayments occur regularly in the Intermediate 2 exam and can be worth between four and six marks. Calculators will be allowed. Look at the example below.

Example

The table shown below is used to calculate loan repayments.

Monthly repayments on a loan of £1000

APR	12 months	24 months	36 months	48 months
12%	£90·61	£48·84	£34·97	£28·08
14%	£91·45	£49·67	£35·83	£28·96
16%	£92·28	£50·51	£36·68	£29·85
18%	£93·10	£51·33	£37·54	£30·73
20%	£93·91	£52·15	£38·39	£31·62

Mary O'Donnell borrows £7500 over 24 months at an annual percentage rate (APR) of 16%. Use the table to calculate the total cost of the loan.

Example continued

(*Solution*) Look up the table for monthly repayment (24 months, 16%) = £50·51.
This payment is for a loan of £1000.
Hence monthly payment for a loan of £7500 = 7·5 × £50·51
= £378·825
Hence total repayments over 24 months = 24 × £378·825
= £9091·80.
Hence cost of Loan = £(9091.80 − 7500) − £1591·80.

This example is worth four marks.

You should not immediately round sums of money such as £378·825. Any rounding required should be left until the end of all the calculations.

Some students would mistakenly do a calculation involving 16%. The percentage is there to help you find the correct amount from the table, i.e. £50·51. It should not be used in any calculation.

If you are asked for the **cost** of the loan, remember to subtract the loan at the end!

Credit Cards

Credit cards can be used to pay for goods and services, and also for getting cash from banks and cash dispensers. Credit card statements are issued monthly. The statement will include details of the minimum payment to be made. If the full balance is not paid off, interest is added the following month.

Example

Here is part of Bob Spender's monthly credit card statement:

Purchases	£
World Garden	15·99
Decking Company	165·85
Bright's Lights	99·99
Balance owed	281·83

Minimum payment £5 or 3%, whichever is greater

Interest rate = 1·67% per month

a) Bob makes the minimum payment. How much does he pay?
b) During the next month, Bob adds purchases of £100 to his credit card.
Calculate the balance owed on his next statement.

Example continued ➤

Example *continued*

(*Solution*)   a)   3% of £281·83 = 0·03 × 281·83 = 8·4549 =£8·45 (to nearest p)
Since £8·45 > £5, the minimum payment is £8·45.

b)   Balance owed =   £281·83
Less payment   −8·45
=   £273·38

Interest = 1·67% of 273·38 = 0·0167 × 273·38 = £4·57.
Purchases = £100

Hence balance owed = £(273·38 + 4·57 + 100)
= £377·95.

This example would be worth five marks, two for part (a), three for part (b).

Note that an alternative to calculating 1·67% of £273·38 and then adding it on would be to multiply 273·38 by 1·0167 leading to 277·95. This method, explained in Chapter 3, increases £273·38 by 1·67% and the £100 purchases can then be added on giving £377·95.

Remember that you should calculate answers to the nearest penny in credit card statements.

LOGIC DIAGRAMS

Four topics will be considered in this outcome – network diagrams, flowcharts, tree diagrams and spreadsheets. In the exam, between one and three of these topics could be tested in a given year, and there could be a spread of between four and eleven marks allocated.

Network Diagrams

A network diagram consists of vertices (or nodes) connected by arcs (or edges).

The degree of a vertex (or node) is the number of arcs meeting there.

Consider the network diagram shown below.

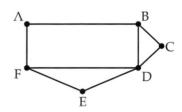

This network has 6 vertices (or nodes). These are represented by the dots at A, B, C, D, E and F. The network has 8 arcs (or edges) connecting the vertices.

The network can be used to plan a route between the vertices.

The degree of each vertex (or node) is given below.

   A – 2, B – 3, C – 2, D – 4, E – 2, F – 3.

A, C, D and E are **even** vertices, while B and F are **odd** vertices.

If a network has more than 2 odd vertices, then the network cannot be covered without going over at least one arc twice.

Example

A network is **traversable** if it can be drawn by going over every line once and only once without lifting your pencil from the paper.

The network shown opposite can be traversed by the route

   B→D→A→B→C→D

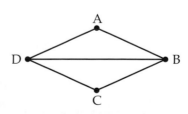

Example continued ➤

Example *continued*

Is the following network traversable?

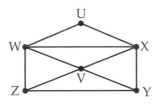

Explain your answer.

(Solution)   Find the degree of the vertices: U – 2, V – 4, W – 4, X – 4, Y – 3, Z – 3.

Yes, the network is traversable as it has only 2 odd vertices (at Y and Z)

or

Yes, it is traversable by the route

$Y \rightarrow Z \rightarrow V \rightarrow Y \rightarrow X \rightarrow V \rightarrow W \rightarrow X \rightarrow U \rightarrow W \rightarrow Z.$

Hints *and* **Tips**

You should start questions like this by writing down the degree of each vertex, and then counting how many odd vertices there are.

Either of the two answers given above is acceptable. You do not have to give both. If you decide to give a pathway through the network, it must start at one odd vertex and finish at the other. In this example, the pathway must start at Y and finish at Z, or vice versa.

This question is worth two marks.

Example

The diagram below shows part of a bus route.

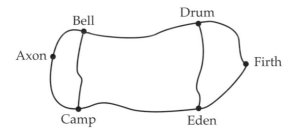

A bus inspector has to travel along every route shown.
Is it possible to do this without travelling any route more than once?
You must give a reason for your answer.

Example *continued* ➣

Example *continued*

(Solution)   Degree of each vertex: Axon – 2, Bell – 3, Camp – 3,
Drum – 3, Eden – 3, Firth – 2.
There are four odd vertices.
It is not possible, as there are more than two odd vertices.

Critical Path

A type of network in which the vertices represent the start and finish of a stage in a job is called an **activity network**. The 'length' of the longest path is the minimum time it will take to complete the job. The longest path is called the **critical path**.

Example

This network shows how three people tidy the sitting room in their house.
All three people work at the same time.
All times are in minutes.

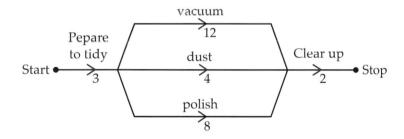

By considering the critical path, find the minimum time required to tidy the sitting room.

(Solution)   The critical path is the longest path = (3 + 12 + 2) minutes.
The minimum time to complete the job is therefore 17 minutes.

There would be one mark for this question. You would not need a calculator for any questions on network diagrams, so they would probably appear in Paper 1.

Tree Diagrams

Tree diagrams are used to list different possibilities when there is a choice of categories; for example, starter, main course, and dessert in a three-course meal.

Example

This mileage chart shows the distances between Scotland's four largest cities.

Aberdeen

147	**Glasgow**		
67	80	**Dundee**	
125	46	58	**Edinburgh**

A courier leaves Aberdeen. He has to deliver parcels to Glasgow, Dundee and Edinburgh. He cannot go through any city more than once and he does not have to return to Aberdeen.

a) Draw a tree diagram to show all possible delivery routes.

b) Which is the shortest route?

(Solution) **a)**

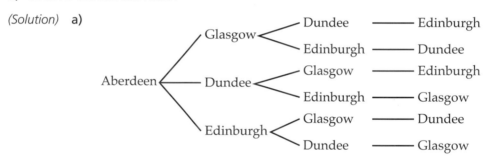

b) A–G–D–E = 147 + 80 + 58 = 285, A–G–E–D = 147 + 46 + 58 = 251.

A–D–G–E = 67 + 80 + 46 = 193, A–D–E–G = 67 + 58 + 46 = 171.

A–E–G–D = 125 + 46 + 80 = 251, A–E–D–G = 125 + 58 + 80 = 263.

Hence Aberdeen–Dundee–Edinburgh–Glasgow is shortest (171 miles).

This question would be worth five marks (three for part (a) and two for part (b)). Because of the calculations in part (b), it would probably appear in Paper 2. While it is straightforward, there are a few issues to be aware of. Some students are unsure of how to read a mileage chart. If you are in that position, then ask for advice. Make sure you can draw tree diagrams, and in part (b), it would be essential for you to show **all** the working as laid out, even though it will take some time.

Flowcharts

You should have met flowcharts often. Remember to start them at the top, answer YES or NO when you come to a decision box, and you should have no problems. Flowchart questions occur fairly often and usually in Paper 2. Such questions are worth three or four marks. Use a calculator in the following example.

Example

Anne Simpson has a job selling encyclopedias. The flowchart below shows how her pay is calculated each month.

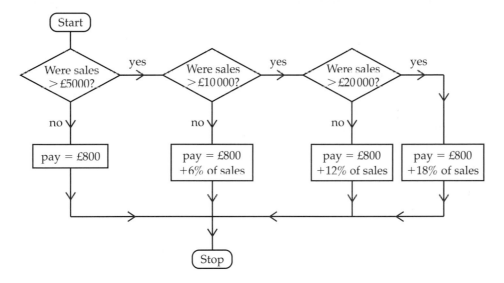

Calculate Anne's pay in a month in which she sells £12 500 worth of encyclopedias.

(Solution) Were sales > £5000?   YES
            Were sales > £10 000?   YES
            Were sales > £20 000?   NO.

            Arrive at box 'Pay = £800 + 12% of sales'.
            Hence pay   = £800 + 12% of £12 500
                             = £800 + 0·12 × £12 500
                             = £800 + £1500
                             = £2300.

Check this working carefully, and practise a few flowcharts. Watch out for the order of operations when doing the calculations (in this case do the multiplication before the addition).

Spreadsheets

You should be able to enter data into a spreadsheet, enter a formula into a spreadsheet and use SUM and AVERAGE in a spreadsheet. Each box in a spreadsheet is called a *cell*.

You are likely to be asked to write down a formula for a cell in a spreadsheet.

Here are a few things to keep in mind.

> ## Key Points
>
> ◆ Use * instead of × for multiplication.
> ◆ Use / instead of ÷ for division.
> ◆ If you are asked for a formula, **always** start with an equals sign (=).
> ◆ If you are asked for a formula which involves adding several cells, e.g. A2 to A7, the formula would be = SUM(A2:A7) or = SUM(A2..A7). You **must** include the equals sign at the beginning, write SUM in capital letters, write A2 and A7 in brackets, with a colon : or two dots .. between A2 and A7 as 'punctuation'.
> ◆ If you are asked to write down a formula which involves finding the average of several cells, e.g. A2 to A7, the formula would be = AVERAGE(A2:A7) or = AVERAGE(A2..A7). Again, remember the equals sign and use capital letters. **Do not** shorten AVERAGE to AVE or AVG.

Example

Mr. Wilson gives his students three tests. He records their marks on a spreadsheet.

	A	B	C	D	E
1		Test 1	Test 2	Test 3	Total
2	Allison	45	53	67	
3	Brown	56	57	55	
4	Charles	78	67	79	
5	Davis	45	38	47	
6	Fielding	50	53	56	
7	Glass	67	67	68	
8	Jones	80	83	90	
9	McDonald	45	54	67	
10	Roberts	32	28	34	
11	Twain	57	63	70	
12					
13	average				

a) Write down the **formula** to enter in cell E2 the total mark for Allison.
b) Write down the **formula** to enter in cell B13 the average mark for Test 1.

(Solution) **a)** = SUM(B2:D2) or = SUM(B2..D2) or = B2+C2+D2.

b) = AVERAGE(B2:B11) or = AVERAGE(B2..B11) or
= SUM(B2:B11)/10 or = SUM(B2..B11)/10 or
= (B2 + B3 + B4 + B5 + B6 + B7 + B8 + B9 + B10 + B11)/10.

There is one mark for each part here. Although alternative correct versions are given in part (b), the first two, using AVERAGE, would be recommended. Always remember the = sign, capital letters, punctuation and brackets, when writing formulae involving SUM or AVERAGE.

Example

Gary Thomson borrows £20 000 from the Blue Shark Loan Company.
He wants to make payments of £800 per month.
Blue Shark Loan company adds 0·65% interest each month.
He designs a spreadsheet to work out how much he owes after interest is added and his
payment is made each month.

	A	B	C	D
1	**Blue Shark Loan Company**			
2				
3	Interest charged		0·65% per month	
4				
5	Amount owed		£20 000	
6	Monthly payment		£800	
7				
8	Amount owed after:		**Interest**	**Payment**
9	January		£20 130·00	£19 330·00
10	Febuary		£19 455·65	£18 655·65
11	March		£18 776·91	£17 976·91
12	April		£18 093·76	£17 293·76
13	May		£17 406·17	£16 606·17
14	June		£16 714·11	£16 014·11
15	July		£16 017·55	£15 217·55
16	August		£15 316·46	£14 516·46
17	September		£14 610·82	£13 810·82
18	October		£13 900·59	£13 100·59
19	November		£13 185·74	£12 385·74
20	December			

a) Write down the **formula** to enter in cell C20 the amount owed in December after interest
has been added.

b) The result of the formula =C20 − 800 is entered in cell D20.
What will appear in cell D20?

(Solution)   a)   =D19*1·0065.

           b) $D20 = C20 - 800 = D19 \times 1·0065 - 800$

$$= £12 385·74 \times 1·0065 - 800$$

$$= £11 666·25$$

The formula in part (a) is tricky. You have to increase the amount in cell D19 by 0·65%. To do
this you must use the method from Chapter 3 on percentages. This reminds you that the new
amount will be (100 + 0·65)% = 100·65% of the old amount. Remember, too, that
100·65% = 1·0065 as a decimal. So you must multiply cell D19 by 1·0065. Remember to use *
for the multiplication sign in a spreadsheet.

This question is worth three marks, (one for part (a) and two for part (b)).

This completes the section on logic diagrams. Now we shall look at formulae.

FORMULAE

 In this outcome you should be able to interpret and use formulae expressed in words or, more commonly, in symbols. This is usually tested in Paper 1, so do not use a calculator. This is tested most years and is worth between three and six marks in total.

It is common to be given a two-part question, with the first part using the formula directly, and the second part asking you to find the value of one variable in the formula which is not the subject of the formula.

Most students can substitute values into formulae very well, but many have difficulty with the calculations, especially if a calculator is not allowed.

We shall look at the best way of doing the calculations. It is important that you carry out calculations in the correct order. Remember that multiplication and/or division are done before addition and/or subtraction. In fact, you may know the memory aid BOMDAS which gives you the order – Brackets, Of, Multiply, Divide, Add, Subtract. Remember it!

Example

The area of a shape is given by the formula:

$$A = \frac{1}{2}(a + b)h.$$

a) Calculate A when $a = 12$, $b = 5$ and $h = 8$.
b) Calculate h when $A = 66$, $a = 16$ and $b = 6$.

(Solution) a) $A = \frac{1}{2}(a + b)h = \frac{1}{2} \times (12 + 5) \times 8 = \frac{1}{2} \times 17 \times 8$

$$= 4 \times 17$$
$$= 68.$$

b) $A = \frac{1}{2}(a + b)h \Rightarrow 66 = \frac{1}{2} \times (16 + 6) \times h$

$$\Rightarrow 66 = \frac{1}{2} \times 22 \times h$$

$$\Rightarrow 66 = 11 \times h$$

$$\Rightarrow h = \frac{66}{11} = 6.$$

 This is worth six marks (three for each part). The fraction makes this problem a little difficult.

In part (a) attend to the brackets first (BOMDAS). Notice also in part (a) that we could have multiplied 17 by 8, then divided 136 by 2 to get the answer 68. However, it is easier to divide 8 by 2 then multiply by 17. Always look for the easiest way to multiply numbers. You should not divide 17 by 2 = 8·5 and then multiply by 8. The answer is still 68 but the calculation is much harder.

In part (b), it is easier to deal with the fraction quickly by dividing 22 by 2 to give 11. Life is a lot simpler once fractions are out of the way!

I would strongly advise you not to change the subject of the formula in part (b). It is more difficult to do this and if you make a mistake, the maximum mark available is one out of three. It is safer to do it as shown.

Example

The surface area, A (square centimetres), of a cone is given by the formula:

$$A = \pi r(r + s),$$

where r (centimetres) is the radius of the base
and s (centimetres) is the slant height.

Take $\pi = 3\cdot14$

Calculate A when $r = 2$ and $s = 8$.

(Solution) $A = \pi r(r + s) = 3\cdot14 \times 2 \times (2 + 8) = 3\cdot14 \times 2 \times 10$
$$= 6\cdot28 \times 10$$
$$= 62\cdot8.$$

This is worth three marks; two for the substitution and one for the calculation. You must replace π by 3·14. Remember to deal with the brackets first and then the multiplication.

Example

The distance, D (metres), travelled by an accelerating object is given by the formula

$$D = ut + \frac{1}{2}ft^2$$

where u (metres per second) is the initial velocity,
            t (seconds) is the time taken,
and f (metres per second per second) is the acceleration.

a) Calculate D when $u = 20$, $t = 6$ and $f = 8$.

b) Calculate f when $D = 88$, $u = 12$ and $t = 4$.

Example *continued* ➤

HOW TO PASS INTERMEDIATE 2 MATHS

Example continued

(Solution) a) $D = ut + \frac{1}{2}ft^2 = 20 \times 6 + \frac{1}{2} \times 8 \times 6^2 = 120 + \frac{1}{2} \times 8 \times 36$

$$= 120 + 4 \times 36$$
$$= 120 + 144$$
$$= 264.$$

b) $D = ut + \frac{1}{2}ft^2 \Rightarrow 88 = 12 \times 4 + \frac{1}{2} \times f \times 4^2$

$$\Rightarrow 88 = 48 + \frac{1}{2} \times f \times 16$$
$$\Rightarrow 88 = 48 + 8 \times f$$
$$\Rightarrow 8 \times f = 88 - 48 = 40$$
$$\Rightarrow f = \frac{40}{8} = 5.$$

Check all the working carefully, note the order of operations, then practise similar examples from your textbook without using a calculator.

Chapter 18

FURTHER STATISTICS

This is the final outcome in the Applications Unit. You are required to do a *statistical assignment* **during your course and this should also help to prepare you for exam questions.**

For your assignment, you are expected to compare data from two similar sources by constructing tables, drawing graphs and analysing the data through calculation of averages and measures of spread. The assignment does not count towards your final exam mark, however.

There is usually a question in the exam from this outcome worth five or six marks. It involves either finding the mean from a frequency table, or drawing a cumulative frequency diagram.

We shall have a detailed look at both later. Firstly, however, we shall look at *histograms* and how to estimate the *mode*.

Finding the Mode from a Histogram

Suppose you are asked to estimate the value of the mode from the following histogram which shows the heights, h centimetres, of a group of students.

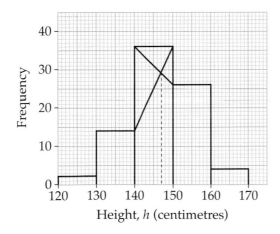

Height, h (centimetres)

The **modal class** is the class interval with the highest frequency, and here it is the class interval between 140 and 150 (normally written as $140 \leqslant h < 150$). To estimate the value of the mode, you should draw a cross on the histogram, as shown, by joining the top corners of the modal class to the tops of the intervals on either side.

The mode is found by reading the value on the x-axis below the point of intersection.

In this example, the estimate of the mode is 147 cm.

Data which has been *measured* is called **continuous** data, whereas data which has been *counted* is called **discrete** data.

Class intervals such as 0–4, 5–9, 10–14, etc. are usually used for *discrete* data. This type of data is often shown in a bar graph.

Class intervals of the type $0 \leqslant h < 5$, $5 \leqslant h < 10$, $10 \leqslant h < 15$, are used for *continuous* data. This type of data is often shown in a histogram such as the one above. Be aware that there should be a continuous scale on the *x*-axis when drawing a histogram.

Important as the last example is, the next section on finding the mean from a frequency table using midpoints is possibly more important. It occurs in the exam regularly and is worth between five and six marks.

Finding the Mean From a Frequency Table

Example) The marks of a group of students in their Intermediate 2 exam, out of 80, are listed below.

$$40 \quad 45 \quad 38 \quad 57 \quad 36 \quad 32 \quad 61 \quad 55 \quad 56 \quad 48$$
$$56 \quad 45 \quad 47 \quad 49 \quad 60 \quad 39 \quad 42 \quad 49 \quad 56 \quad 59$$
$$40 \quad 50 \quad 60 \quad 37 \quad 54 \quad 54 \quad 46 \quad 63 \quad 45 \quad 57.$$

a) Construct a frequency table with class intervals

30–34, 35–39, 40–44, etc.

b) Calculate the mean mark.

(Solution) In part (a), you will have to extend the class intervals well beyond 40–44. Look for the highest number in the list, 63. This means that the final class interval will be 60–64. I would advise that you go through each number in the list individually, using tally marks to complete the frequency table. Always check by counting that you have not missed out any numbers. There are 30 numbers in the list, so the frequencies should add up to 30.

Take time and care with part (a). It is easy to lose a mark through carelessness.

In readiness for part (b), leave room for columns for 'midpoints' and '*fx*' (midpoint × frequency). Hence:

a) class interval	midpoint	tally	frequency, *f*	*fx*
30–34	32	|	1	$32 \times 1 = 32$
35–39	37	||||	4	$37 \times 4 = 148$
40–44	42	|||	3	$42 \times 3 = 126$
45–49	47	|||| |||	8	$47 \times 8 = 376$
50–54	52	|||	3	$52 \times 3 = 156$
55–59	57	|||| ||	7	$57 \times 7 = 399$
60–64	62	||||	4	$62 \times 4 = 248$
			$\Sigma f = 30$	$\Sigma fx = 1485$

b) Mean $= \dfrac{\Sigma fx}{\Sigma f} = \dfrac{1485}{30} = 49{\cdot}5$

To find the mean, you need *midpoints*. Here it is easy as 32 is clearly midway between 30 and 34. It is not always so obvious. Note that $(30 + 34) \div 2 = 32$, so to find midpoints, add the endpoints of the interval and divide by 2.

Always include the *fx* column, found by multiplying the midpoint value by the frequency.

Always add the frequency column to get Σf, the sum of the frequencies, **and** always add the *fx* column to get Σfx, the sum of the midpoint values $\times$ the frequencies.

You should **not** add the midpoint column.

Then use the formula, mean $= \dfrac{\Sigma fx}{\Sigma f}$ to complete the question.

The process of finding the mean takes time and patience, but is not difficult as long as you have prepared and know what you are doing. So be ready! This type of question crops up more often than not. Practise the techniques until you are confident. Often the frequency table has already been done for you so you may not have to do part (a).

For Practice

Try and find the mean mark from this data for practice.

Exam marks	Frequency
10–19	2
20–29	4
30–39	7
40–49	10
50–59	22
60–69	48
70–79	5
80–89	2

There is no need for tally marks here as the frequencies are given. Add a column for midpoints and an *fx* column, then find the mean. Check that your answer is 56·5.

Well done if you got it right, for this example is worth five marks.

Now we shall look at *cumulative frequency*.

Cumulative Frequency Diagrams

Example) The speeds of 100 cars travelling along a stretch of road were recorded. The results are shown in the table below.

Speed (s km/h)	Frequency
$10 \leqslant s < 20$	4
$20 \leqslant s < 30$	12
$30 \leqslant s < 40$	14
$40 \leqslant s < 50$	28
$50 \leqslant s < 60$	32
$60 \leqslant s < 70$	10

a) Construct a cumulative frequency column for the above data.
b) Using squared paper, draw a cumulative frequency diagram for this data.
c) From your diagram, estimate the median speed of a car.

This is worth five marks (one for parts (a) and (c), and three for part (b)). We have already looked at how to construct a cumulative frequency column in Chapter 10. I hope you remember. Part (b) is most likely to cause problems. (A cumulative frequency diagram or curve is also called an *ogive*.)

(Solution) a)

Speed (s km/h)	Frequency	Cumulative frequency
$10 \leqslant s < 20$	4	4
$20 \leqslant s < 30$	12	16
$30 \leqslant s < 40$	14	30
$40 \leqslant s < 50$	28	58
$50 \leqslant s < 60$	32	90
$60 \leqslant s < 70$	10	100

b)

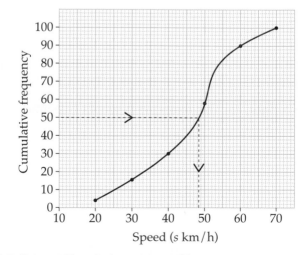

c) Median = 48 km/h (approximately).

In part (a), did you remember how to fill in the cumulative frequency column? If you are running short of time in your exam, you do not need to re-write the whole frequency table, it would be acceptable to write 4, 16, 30, 58, 90, 100. If you do this, however, be extra careful when plotting points for the cumulative frequency diagram.

Part (b) is usually poorly done in exams, so I will now explain in detail how to draw a cumulative frequency diagram, such as the one above.

Key Points

◆ Firstly, the cumulative frequency axis should **always** be vertical, therefore the speeds will be horizontal. Clearly label the axes.

◆ The scales cause most problems. The horizontal scale in a cumulative frequency diagram must always be continuous as shown, i.e. 10, 20, 30, etc. spaced equally along the x-axis. **Never** mark class intervals along the x-axis. You will lose marks if you mark $10 \leqslant s < 20$, $20 \leqslant s < 30$, on the axis.

◆ The vertical scale is straightforward, extending as far as the final entry in the cumulative frequency column.

◆ Plotting the points causes problems to some students. Always plot the cumulative frequency against the **upper** limit of the class interval. This means that, in this example, the points plotted were (20, 4), (30, 16), (40, 30), (50, 58), (60, 90) and (70, 100). Plot all the points accurately. To help with this, choose a *suitable* scale on your squared paper.

◆ Finally, when drawing the graph, draw as *smooth* a curve as you can, making sure it passes through all the points. Do not merely 'join the dots'.

In part (c) of the last example, we estimate the median by drawing a horizontal line along from 50 ($\frac{1}{2}$ of 100) on the vertical axis to the curve, and drawing down a vertical line to the x-axis to read off the answer.

You could also be asked for the upper and lower quartiles from a cumulative frequency diagram. In the last example, draw along from 25 $\left(\frac{1}{4} \text{ of } 100\right)$ for the lower quartile and along from 75 $\left(\frac{3}{4} \text{ of } 100\right)$ for the upper quartile. It would then be possible to find the semi-interquartile range.

There is a lot for you to take in here, but it is not too difficult once you have had some practice. Try several cumulative frequency diagrams from your textbook and your confidence will soon increase. The topic appears regularly in the exam.

This brings us to the end of this outcome and also to the end of the syllabus.

In the following Appendix you will find a typical Paper 1 to practise.

APPENDIX 1: PRACTICE PAPER 1

Now you can try a practice exam paper – Paper 1 (Non-calculator). You should allow 45 minutes to do this paper. The standard of the questions is the same as in the actual exam, and the paper is of the same length. It is worth 30 marks.

All students should do Section A, covering Units 1 and 2.

Do section B **only** if you are studying Unit 3.

Do section C **only** if you are studying Unit 4, the Applications of Mathematics.

Once you have finished you will find detailed solutions at the end.

Paper 1
Section A – All students do this section

1

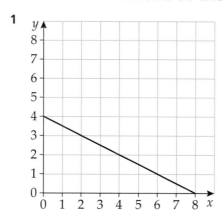

Find the equation of the straight line. **(3)**

2 a) Multiply out the brackets and collect like terms:

$$(3x + 2)(x^2 - x + 5).$$ **(3)**

b) Factorise:

$$3x^2 + 5x - 2.$$ **(2)**

3 Alan counts how many matches there are in a sample of matchboxes, with these results:

48 50 47 48 51 47 50 49 49 50
49 46 45 47 50 48 49 50 51 49.

a) Construct a frequency table from the above data and add a cumulative frequency column. **(2)**

b) What is the probability that a matchbox chosen at random from this sample has more than 48 matches? **(1)**

4 A group of girls was asked how many portions of fruit they had eaten one week. The results are listed below:

     7 12 5 3 15 9 6 2 1 5 4 8 12.

a) For the given data, calculate:
  (1) the median;   **(1)**
  (2) the lower quartile;   **(1)**
  (3) the upper quartile.   **(1)**

b) Draw a boxplot to illustrate this data.   **(2)**

A group of boys was asked how many portions of fruit they had eaten that week. Their results are shown in the boxplot below.

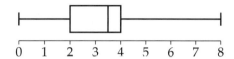

c) By comparing the boxplots, make **two** appropriate comments about the numbers of portions of fruit eaten by the girls and boys.   **(2)**

5 The square below has length $(2x - 3)$ centimetres.

$(2x - 3)$ cm

If the area of the square is 49 square centimetres, find x.   **(2)**

Section B – Only do this section if you are studying Unit 3.

6

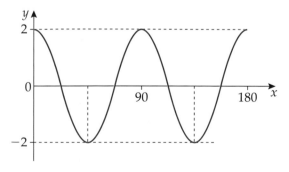

The graph of $y = a \cos bx°$ is shown in the diagram.
State the values of a and b.   **(2)**

7

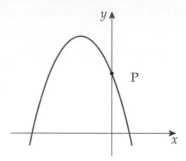

The equation of the parabola in the above diagram is
$$y = 10 - (x + 1)^2.$$

a) State the coordinates of the maximum turning point of the parabola. **(2)**
b) State the equation of the axis of symmetry. **(1)**
c) Find the coordinates of P. **(2)**

8 Express $\sqrt{8} - \sqrt{2} + \sqrt{50}$ as a surd in its simplest form. **(3)**

Section C – Only do this section if you are studying Unit 4, the Applications of Mathematics

6 Andrew has completed his first week in his new job.
His basic rate of pay is £4·80 per hour. He is paid time and a half for working overtime.
During the week, he worked 40 hours at the basic rate and 4 hours overtime.
How much did Andrew earn? **(3)**

7 A network diagram is shown below.

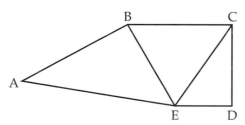

How many **odd** vertices are there in the network diagram? **(1)**

8 The surface area of the shape below is given by the formula:
$$S = 2l^2 + 4lh.$$

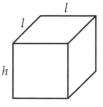

a) Calculate S when $l = 6$ and $h = 4$. **(3)**
b) Calculate h when $S = 210$ and $l = 5$. **(3)**

Paper 1 – Solutions

Section A – All students

1 Gradient, $m = \dfrac{y_2 - y_1}{x_2 - x_1} = \dfrac{4 - 0}{0 - 8} = \dfrac{4}{-8} = -\dfrac{1}{2}$ or $m = \dfrac{\text{vertical}}{\text{horizontal}} = -\dfrac{4}{8} = -\dfrac{1}{2}$.

y-intercept, $c = 4$.

Hence equation of Line is $y = mx + c \implies y = -\dfrac{1}{2}x + 4$.

2 a) $(3x + 2)(x^2 - x + 5) = 3x(x^2 - x + 5) + 2(x^2 - x + 5)$
$= 3x^3 - 3x^2 + 15x + 2x^2 - 2x + 10$
$= 3x^3 - x^2 + 13x + 10$.

b) $3x^2 + 5x - 2 = (3x - 1)(x + 2)$.

3 a)

no. of matches	frequency	cumulative frequency
45	1	1
46	1	2
47	3	5
48	3	8
49	5	13
50	5	18
51	2	20

b) Probability of more than 48 matches $= \dfrac{(5 + 5 + 2)}{20} = \dfrac{12}{20}$.

4 a) Put numbers in order: $1\ 2\ 3\ |\ 4\ 5\ 5\ 6\ 7\ 8\ 9\ |\ 12\ 12\ 15$

There are 13 numbers, so median is in $(13 + 1) \div 2 = 7^{\text{th}}$ position.
(1) Median, $Q_2 = 6$.
(2) Lower Quartile, $Q_1 = 3\cdot5$.
(3) Upper Quartile, $Q_3 = 10\cdot5$.   (Note: $(9 + 12) \div 2 = 10\cdot5$.)

b)

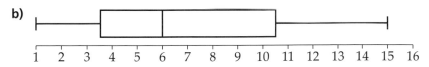

c) The median for the girls is 6, while the median for the boys is only 3·5. This suggests that the girls eat more fruit than the boys.

The range for the boys is 8 $(8 - 0)$, while the range for the girls is 14 $(15 - 1)$.

Therefore the results for the girls are more spread out than for the boys.
(This is only one of many possible answers to part (c). However, any answer should compare both the medians **and** the spread of both sets of data.)

5 Area of square $= 49\,\text{cm}^2 \implies$ Length of side of square $= \sqrt{49} = 7$ cm.

Hence $2x - 3 = 7 \implies 2x = 7 + 3 \implies 2x = 10 \implies x = 5$.

(You could also start this question by saying that $(2x - 3)^2 = 49$.

This method would lead to a more difficult equation based on Unit 3 work.

So you should be alert. As the question is worth only 2 marks, there is an easier way to tackle it.)

Section B – Unit 3 students only

6 $a = 2$, $b = 4$.

7 **a)** $(-1, 10)$.

b) $x = -1$.

c) P lies on the y-axis.
The parabola cuts the y-axis where $x = 0$.
Substitute $x = 0$ into $y = 10 - (x + 1)^2 \Rightarrow y = 10 - (0 + 1)^2$
$$\Rightarrow y = 10 - 1^2 = 9.$$
Coordinates of P are $(0, 9)$.

8 $\sqrt{8} - \sqrt{2} + \sqrt{50} = \sqrt{4 \times 2} - \sqrt{2} + \sqrt{25 \times 2} = 2\sqrt{2} - \sqrt{2} + 5\sqrt{2}$
$$= (2 - 1 + 5)\sqrt{2}$$
$$= 6\sqrt{2}.$$

Section C – Applications of Mathematics students only

6 Pay at Basic Rate $= 40 \times 4{\cdot}80 = 4 \times 48 = £192$.
Overtime Rate of Pay $= 4{\cdot}80 + 2{\cdot}40 = £7{\cdot}20$.
Pay at Overtime Rate $= 4 \times 7{\cdot}20 = £28{\cdot}80$.

Total Amount earned $= 192{\cdot}00 + 28{\cdot}80 = £220{\cdot}80$.

7 The degree of each vertex is: A–2, B–3, C–3, D–2, E–4.
Therefore there are **2** odd vertices.

8 **a)** $S = 2l^2 + 4lh = 2 \times 6^2 + 4 \times 6 \times 4 = 2 \times 36 + 96 = 72 + 96 = 168$.

b) $S = 2l^2 + 4lh \Rightarrow 210 = 2 \times 5^2 + 4 \times 5 \times h \Rightarrow 210 = 2 \times 25 + 20 \times h$
$$\Rightarrow 210 = 50 + 20 \times h$$
$$\Rightarrow 20 \times h = 210 - 50 = 160$$
$$\Rightarrow h = \frac{160}{20} = 8.$$

APPENDIX 2:
PRACTICE PAPER 2

Now you can try another practice exam paper – Paper 2 (Calculator). Allow 1 hour 30 minutes to do this paper. It is worth 50 marks. The level of difficulty is similar to that in your actual exam.

All students should do Section A, covering Units 1 and 2.

Do Section B **only** if you are studying Unit 3.

Do Section C **only** if you are studying Unit 4, the Applications of Mathematics.

There are detailed solutions at the end.

Paper 2
Section A – All students do this section.

1   After a survey of prices, it is predicted that the price of goods bought in a supermarket will increase by 2·8% per year.
    At present, the supermarket price of a carton of milk is £1·12.
    Calculate the predicted price of this carton in 3 years.
    Give your answer to the nearest penny.   **(3)**

2   **a)** The number of toffees in six different packets is counted.
        The results are shown below:

            23   24   21   25   22   29.

    Use an appropriate formula to calculate the mean and the standard deviation of these results.
    Show clearly all your working.   **(4)**

    The manufacturers intend that the mean number of toffees per packet should be 25(±2) and that the standard deviation should be less than 2.

    b) Have the manufacturers' intentions been met?
        Explain clearly your answer.   **(2)**

3   O is the centre of two concentric circles.
    PQ is a tangent to the smaller circle and a chord of the larger circle.
    The radius of the larger circle is 15 centimetres.
    The radius of the smaller circle is 10 centimetres.
    Calculate the length of PQ.   **(4)**

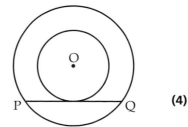

4   Seats on the Paris to Lyon train are sold to adults and children.
An adult ticket costs 55 Euros, and a child ticket costs 28 Euros.
On one journey a total of 145 tickets were sold.
Let a be the number of adult tickets sold and c be the number of child tickets sold.

a) Write down an equation in a and c which satisfies the above condition.   **(1)**

The sale of the tickets on this journey totalled 6598 Euros.

b) Write down a second equation in a and c which satisfies this condition.   **(1)**

c) How many adult tickets and how many child tickets were sold?   **(4)**

5   A farmer makes blocks of cheese in the shape of hemispheres, with radius 8 centimetres.

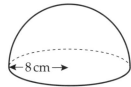

a) Calculate the volume of **one** block of cheese.
Give your answer correct to two significant figures.   **(3)**

He also makes blocks of cheese in the shape of cylinders.

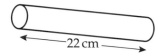

b) The cylindrical block of cheese has the **same** volume as the hemispherical
one. It has length 22 centimetres.
Calculate the radius of this block.   **(3)**

6   A flagpole is held in position by two wires.

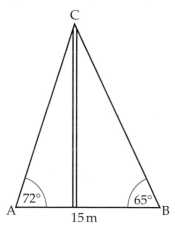

The wires stretch from two points on the ground, A and B, to the top of the flagpole at C.
The wire at A makes an angle of 72° with the ground.
The wire at B makes an angle of 65° with the ground.
The distance AB is 15 metres.

Find the height of the flagpole.   **(5)**

7 The diagram below shows a sector of a circle.

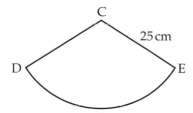

C is the centre of the circle.
D and E are points on the circumference of the circle.
The radius of the circle is 25 centimetres.
The area of the sector is 550 square centimetres.
Find the size of angle DCE. **(4)**

Section B – Only do this section if you are studying Unit 3.

8 Solve the equation

$$3x^2 + 5x - 4 = 0,$$

giving your answers correct to one decimal place. **(4)**

9 a) Simplify $\dfrac{(x-3)^2}{5x-15}$. **(2)**

  b) Simplify $3m^{\frac{5}{3}} \times 2m^{\frac{1}{3}}$. **(2)**

10 Change the subject of the formula:

$$h = Kn^2 - 7 \text{ to } n.$$ **(3)**

11 a) Solve the equation:

$$4\cos x° - 3 = 0 \ (0 \leqslant x < 360).$$ **(3)**

  b) Prove that

$$2\cos^2 x° + 3\sin^2 x° - 2 = \sin^2 x°.$$ **(2)**

Section C – Only do this section if you are studying Unit 4, The Applications of Mathematics.

8 The tables below show the monthly repayments to be made, with and without payment protection insurance, when £1000 is borrowed from the Capone Loan Company.

With Payment Protection Insurance

APR	12 months	24 months	36 months	48 months
10%	£93·66	£50·34	£36·03	£28·73
12%	£94·56	£51·27	£37·00	£29·74
14%	£95·46	£52·19	£37·97	£30·75
16%	£96·34	£53·10	£38·94	£31·77

Without Payment Protection Insurance

APR	12 months	24 months	36 months	48 months
10%	£87·52	£45·75	£31·87	£24·96
12%	£88·36	£46·69	£32·72	£25·83
14%	£89·20	£47·42	£33·58	£26·71
16%	£90·03	£48·26	£34·43	£27·60

Sophie wants to borrow £7500. She wants to make repayments over 36 months at an APR of 12% **without** payment protection insurance.
Find the cost of her loan. **(4)**

9 Sam Murphy earns £46 250 per year and has tax allowances totalling £6035.

a) Calculate Sam's taxable income.

b) The rates of tax applicable are as follows:

TAXABLE INCOME	RATE
On the first £34 800	20%
On any income over £34 800	40%

Calculate the amount of tax payable by Sam. **(4)**

10 The flowchart below shows how to calculate the cost of joining a swimming club.

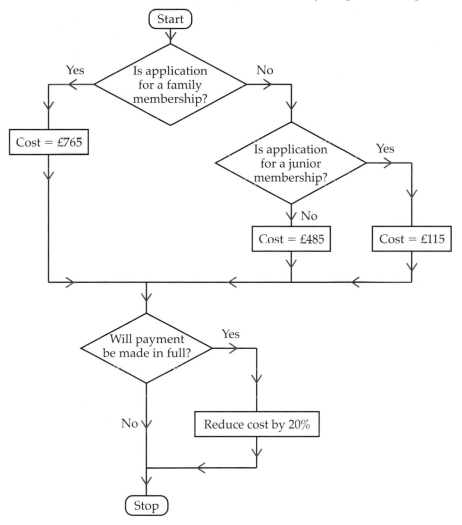

Use the flowchart to calculate the cost for a senior citizen who wants to make the payment in full. **(3)**

11 A survey was carried out to find the delivery time of its pizzas by the 'Pizza the Action' Home Delivery Company.
The results are shown below:

Time in minutes	Frequency
0–4	2
5–9	15
10–14	38
15–19	25
20–24	10
25–29	8
30–34	2

Calculate the mean delivery time in minutes. **(5)**

Paper 2 – Solutions

Section A – All students

1 To increase by 2·8%, multiply by 1·028.
Present price = £1·12.
Price in 3 years = £1·12 × 1·028^3 = £1·22 (to the nearest penny).

2 **a)** Mean, $\bar{x}$ = (23 + 24 + 21 + 25 + 22 + 29) ÷ 6 = 144 ÷ 6 = 24.

Standard deviation, s

Method 1		Method 2	
$s = \sqrt{\dfrac{\Sigma(x - \bar{x})^2}{n - 1}}$		$s = \sqrt{\dfrac{\Sigma x^2 - (\Sigma x)^2 / n}{n - 1}}$	

x	$(x - \bar{x})$	$(x - \bar{x})^2$	x	x^2
23	−1	1	23	529
24	0	0	24	576
21	−3	9	21	441
25	1	1	25	625
22	−2	4	22	484
29	5	25	29	841
		40	144	3496

$$s = \sqrt{\frac{40}{6-1}} = \sqrt{\frac{40}{5}} = \sqrt{8}$$
$$= 2\cdot83$$

$$s = \sqrt{\frac{3496 - 144^2 \div 6}{6-1}} = \sqrt{\frac{40}{5}} = \sqrt{8}$$
$$= 2\cdot83$$

b) The manufacturers' intentions have not been met. Although the mean 24, is within the 25 (±2) range, i.e. between 23 and 27, the standard deviation is 2·83. This is too high as the intention was that it should be less than 2.

3 Use Pythagoras' Theorem

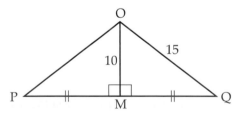

$$OQ^2 = OM^2 + MQ^2$$
$$\Rightarrow \quad 15^2 = 10^2 + MQ^2$$
$$\Rightarrow \quad MQ^2 = 15^2 - 10^2 = 225 - 100 = 125$$
$$\Rightarrow \quad MQ = \sqrt{125}$$
$$= 11\cdot2.$$
$$\Rightarrow \quad PQ = 2 \times MQ = 2 \times 11\cdot2 = 22\cdot4 \text{ cm}.$$

4 **a)** As 145 tickets were sold, the first equation is

$$a + c = 145.$$

b) $55a + 28c = 6598.$

c)

$$a + c = 145 \quad \textbf{(1)}$$
$$55a + 28c = 6598 \quad \textbf{(2)}$$

(1) $\times 28$: $\quad 28a + 28c = 4060 \quad \textbf{(3)}$

(2) − (3): $\quad\quad\quad 27a = 2538$

$$\Rightarrow \quad\quad\quad a = \frac{2538}{27} = 94.$$

substitute $a = 94$ into equation **(1)** $a + c = 145$

$$\Rightarrow \quad\quad 94 + c = 145$$
$$\Rightarrow \quad\quad\quad c = 145 - 94$$
$$\Rightarrow \quad\quad\quad c = 51.$$

(check: $94 \times 55 + 51 \times 28 - 6598$)

There were 94 adult tickets and 51 child tickets sold.
(Remember to answer in words at the end of the question).

5 a) Volume of hemisphere $= \frac{4}{3}\pi r^3 \div 2 = \frac{4}{3} \times \pi \times 8^3 \div 2$

$$= 1072 \cdot 33$$
$$= 1100 \text{ cm}^3 \text{ (to two significant figures)}.$$

b) Volume of cylinder = volume of hemisphere

Hence $\quad \pi r^2 h = 1072$

$\Rightarrow \pi \times r^2 \times 22 = 1072$

$\Rightarrow r^2 = \dfrac{1072}{(\pi \times 22)} = \dfrac{1072}{69 \cdot 1} = 15 \cdot 5$

$\Rightarrow r = \sqrt{15.5}$

$\Rightarrow r = 3 \cdot 9 \text{ cm}$

6 (Use the Sine Rule in triangle ABC.)

Angle ACB $= (180 - 72 - 65)°$
$\quad\quad\quad\quad = 43°$

(The strategy is now to calculate either AC or BC using the Sine Rule, then to use right angled trigonometry (SOH CAH TOA) to calculate the height of the flagpole.)

$$\frac{a}{\sin A} = \frac{b}{\sin B} = \frac{c}{\sin C} \Rightarrow \frac{a}{\sin 72°} = \frac{b}{\sin 65°} = \frac{15}{\sin 43°}.$$

Calculate **either** a: $a = \dfrac{15 \times \sin 72°}{\sin 43°}$ **or** b: $b = \dfrac{15 \times \sin 65°}{\sin 43°}$

$$= 20 \cdot 9 \text{ m} \quad\quad\quad\quad = 19 \cdot 9 \text{ m}$$

Now **either** $\sin 72° = \dfrac{h}{19 \cdot 9}$ **or** $\sin 65° = \dfrac{h}{20 \cdot 9}$.

$\Rightarrow \quad h = 19 \cdot 9 \times \sin 72°$ **or** $h = 20 \cdot 9 \times \sin 65°$

$\Rightarrow \quad h = 18 \cdot 9 \text{ m}.$

The flagpole is therefore 18·9 metres high.

7 Area of circle $= \pi r^2 = \pi \times 25^2 = 1963 \text{ cm}^2$

Hence Angle DCE $= \dfrac{550}{1963} \times 360 = 101°$ (to the nearest degree).

Section B – Unit 3 students only

8 $3x^2 + 5x - 4 = 0 \implies a = 3, b = 5, c = -4.$

Use quadratic formula, $x = \dfrac{-b \pm \sqrt{(b^2 - 4ac)}}{2a} = \dfrac{-5 \pm \sqrt{(5^2 - 4 \times 3 \times -4)}}{2 \times 3}$

$$\implies x = \frac{-5 \pm \sqrt{(25 + 48)}}{6}$$

$$\implies x = \frac{-5 + \sqrt{73}}{6} \ \text{or} \ \frac{-5 - \sqrt{73}}{6}$$

$$\implies x = \frac{3 \cdot 54}{6} \ \text{or} \ \frac{-13 \cdot 54}{6}$$

$$\implies x = 0 \cdot 6 \ \text{or} \ -2 \cdot 3.$$

9 a) $\dfrac{(x-3)^2}{5x-15} = \dfrac{(x-3)^2}{5(x-3)} = \dfrac{(x-3)(x-3)}{5(x-3)} = \dfrac{x-3}{5}.$

b) $3m^{\frac{5}{3}} \times 2m^{\frac{1}{3}} = 3 \times 2 \times m^{\frac{5}{3}} \times m^{\frac{1}{3}} = 6 \times m^{\frac{5}{3} + \frac{1}{3}}$

$$= 6m^{\frac{6}{3}}$$

$$= 6m^2.$$

10 $h = Kn^2 - 7$

$$\implies Kn^2 - 7 = h$$

$$\implies Kn^2 = h + 7$$

$$\implies n^2 = \frac{h+7}{K}$$

$$\implies n = \sqrt{\frac{h+7}{K}}.$$

11 a) $4 \cos x° - 3 = 0$

$$\implies 4 \cos x° = 3$$

$$\implies \cos x° = \frac{3}{4} = 0 \cdot 75.$$

Related angle: Now use $\cos^{-1} 0 \cdot 75 = 41 \cdot 4°$.

Now use the table to find the correct quadrants.

The **Cosine** is **positive** in 1st, 4th quadrants.

Hence $x = 41 \cdot 4$ or $(360 - 41 \cdot 4)$

$= 41 \cdot 4$ or $318 \cdot 6$.

SIN	ALL
$180 - A$	A
TAN	COS
$180 + A$	$360 - A$

b) Hint: Look at the left hand side for a formula you know.

The only possibilities are $\cos^2 x° = 1 - \sin^2 x°$ or $\sin^2 x° = 1 - \cos^2 x°$. Since the right hand side has a $\sin^2$ term, it would be sensible to keep all the $\sin^2$ terms in the working, so substitute $\cos^2 x° = 1 - \sin^2 x°$ into the left hand side and simplify (with your fingers crossed).

Left hand side $= 2 \cos^2 x° + 3 \sin^2 x° - 2$

$= 2 (1 - \sin^2 x°) + 3 \sin^2 x° - 2$

$= 2 - 2 \sin^2 x° + 3 \sin^2 x° - 2$

$= \sin^2 x°$

$=$ Right hand side.

Section C – Applications of Mathematics students only

8 Look up monthly repayment table (36 months, 12%, without PP) = £32·72
This figure is for a loan of £1000.
Hence monthly payment for a loan of £7500 = 7·5 × 32·72 = £245·40.
Total repayments over 36 months = 36 × £245·40 = £8834·40
Hence cost of loan = £(8834·40 − 7500) = £1334·40.

9 a) Taxable Income = Income − Allowances
$$= £(46\,250 - 6035)$$
$$= £40\,215.$$

b) Tax paid at 20% rate = 20% of £34 800 = 0.2 × 34 800 = £6960.
Tax paid at 40% rate = 40% of £(40 215 − 34 800).
$$= 40\% \text{ of } £5415$$
$$= 0·4 \times 5415$$
$$= £2166.$$

Hence total tax paid = £(6960 + 2166)
$$= £9126.$$

10 Is application for a family membership? NO.
Is application for a junior membership? NO.
Arrive at box: Cost = £485.
Will payment be made in full? YES.
Arrive at box: Reduce cost by 20%.
20% of £485 = 0·2 × 485 = £97.
Hence cost = £(485 − 97) = £388.

11

Time in minutes	midpoint	frequency f	fx	
0–4	2	2	2 × 2 =	4
5–9	7	15	7 × 15 =	105
10–14	12	38	12 × 38 =	456
15–19	17	25	17 × 25 =	425
20–24	22	10	22 × 10 =	220
25–29	27	8	27 × 8 =	216
30–34	32	2	32 × 2 =	64
		$\Sigma f = 100$	$\Sigma fx = 1490$	

$$\text{Mean} = \frac{\Sigma fx}{\Sigma f} = \frac{1490}{100} = 14·9.$$

APPENDIX 3:
USEFUL INFORMATION

Sitting the Exam

Make sure that you have the proper equipment with you for the exam. You should write in blue or black ink. Never use red ink or fluorescent pens. You should also bring a pencil and rubber. It is a good idea to use a pencil for drawing graphs such as a cumulative frequency curve in the Applications of Mathematics Unit. It is also essential that you bring compasses for drawing circles, and a protractor in case you are asked to draw a pie chart. You will need a ruler and, of course, a calculator for Paper 2. Make sure it is a calculator you are used to, and check that it is in degree mode. It would be sensible to bring spares for the above.

Make sure you arrive with plenty time to spare before the start of the exam. You don't want to arrive, harassed, at the last minute.

I would recommend that you work through the questions in the order that they appear in the exam. This order has been thought out in advance with the easier questions earlier in the paper. In my experience, students who jump about the paper often get muddled and occasionally forget to do a question!

Please do not leave the exam early before the allocated time is up. While you are still there you can *improve* your mark. It is very unlikely you will have scored 100%. I have known students leave the exam early because their pal left! It doesn't make sense. Stay for the full time and check and re-check your working.

Look at the marks for each question. They will give some idea of how much work is required. If a question is worth only one mark, you will require very little working, if any. For questions worth more than one mark, you **must** show working, or you risk getting no marks. If you make a mistake, score it out. Anything scored out is not marked. However, only score something out if you have something ready to replace it with. It is sad, when marking exams, to see students score out working worth some marks and failing to put anything in its place. However, never leave two answers to the same question.

Take care in copying down information from the question paper and when copying formulae from the formulae sheet. Some students are careless here and it can cost marks.

Similarly, be careful with basic calculations, such as adding and subtracting in your head, using tally marks, listing numbers in order to find the median. Many marks are lost through careless avoidable errors each year. Remember, the more you *practise*, the less likely you are to make such mistakes.

You will write your answers in a booklet with plain white pages. Loose sheets of graph paper are provided for any complicated graphs or diagrams, if required. If you do use graph paper, make sure you put your name on it and put it inside the booklet with your answers at the end of the exam.

It may seem unlikely, but *very* occasionally students sit the *wrong* exam. For example a student sitting Units 1, 2 and 3 might be given the paper for Units 1, 2 and Applications of Mathematics by mistake. This shouldn't happen and is very unlikely, but always check that you have been given the correct exam paper at the start, to be on the safe side.

Vocabulary Index

It is important that you understand all the mathematical words and phrases which might appear in the exam. Scan through the list below, Unit by Unit, outcome by outcome, and make sure you know what they mean. (Each outcome also has its Chapter number indicated.)

(Unit 1)
Chap. 3 (Percentages)
Annum
Appreciation
Compound Interest
Depreciation

Chap. 4 (Volume)
Cone
Cylinder
Hemisphere
Prism
Significant figures
Sphere

Chap. 5 (Straight Line)
Gradient
y-intercept

Chap. 6 (Algebra)
Common factor
Difference of two squares
Expression
Factorise
Term
Trinomial

Chap. 7 (The Circle)
Arc
Bisect
Circumference
Diameter
Perpendicular
Radius
Sector
Tangent

(Unit 2)
Chap. 8 (Trigonometry)
Acute
Area of a triangle
Cosine rule
Quadrant
Related angle
Sine rule
Three-figure bearings

Chap. 9 (Simultaneous Equations)
Coefficient
Substitute

Chap. 10 (Graphs, Charts and Tables)
Boxplot
Cumulative frequency
Data
Distribution
Dotplot
Five-figure summary
Mean
Median
Mode
Pie chart
Quartiles
Semi-interquartile range
Skewed

Chap. 11 (Statistics)

Consistent
Line of best fit
Probability
Scattergraph
Spread
Standard deviation

(Unit 3)
Chap. 12 (Further Algebra)

Cancel
Change the subject of a formula
Denominator
Express
Evaluate
Index (plural: indices)
Numerator
Power
Rational denominator
Simplify
Simplest form
Surd

Chap. 13 (Quadratic Functions)

Axis of symmetry
Decimal places
Maximum turning point
Minimum turning point
Parabola
Quadratic equation
Quadratic formula
Roots

Chap. 14 (Further Trigonometry)

Amplitude
Cosine graph
Identity
Period
Phase angle

Sine graph
Tangent graph
Trigonometric equation

(Unit 4 Applications)
Chap. 15 (Calculations with Money)

Allowances
Balance
Bonus
Commission
Cost of a loan
Credit card
Gross pay
Minimum payment
Net pay
Overtime
Payslip
Repayment
Time and a half

Chap. 16 (Logic Diagrams)

Cell
Critical path
Decision box
Flowchart
Network diagram
Node
Spreadsheet
Traversable
Vertex (plural: vertices)

Chap. 17 (Formulae)

Variable

Chap. 18 (Further Statistics)

Class intervals
Cumulative frequency graph
Histogram
Midpoints
Ogive

Content of Recent Papers

The following charts give an indication of the composition of each Intermediate 2 exam since the year 2000. By studying them, you will get an indication of the frequency with which particular topics have appeared, and the marks they have been worth. Do not assume, however, that because a particular topic has appeared every year or nearly every year, it will definitely appear this year.

Nevertheless the charts should provide some guidance to you in your studying.

UNIT 1

YEAR / PAPER	2000		2001		2002		2002		2003		2004		2005		2006		2007		2008	
	1	2	1	2	1	2	1	2	1	2	1	2	1	2	1	2	1	2	1	2
Percentages, apprecn/deprecn	4			3	4		5				3		3		4		3			4
Volume	7			7	4		5		2	7	5		6		5	2	5			7
$y = mx + c$ (from graph)	3			3	4		3			3	3		5		4		3			4
Straight line (cuts axis)														2	3		2		1	
Simple brackets		3			3				2		3	3								3
Harder brackets				3		3	3								3		3			
Common factor								1										1		
Difference of 2 squares	2															2			1	3
Trinomial factorisation			2			2		2	2			2	2							
Algebra problem			2									3	3							
Length of arc		5		5				4		4	3					5		3		
Area of sector						4								3						
Angle problems in a circle	3				3					3	3							2	3	
Pythagoras/Symmetry in circle		4						4	3					4		4		3		5

USEFUL INFORMATION

UNIT 2

YEAR / PAPER	2000		2001		2002		2002		2003		2004		2005		2006		2007		2008	
PAPER	1	2	1	2	1	2	1	2	1	2	1	2	1	2	1	2	1	2	1	2
Trig angles > 90°							2						1		2		1			
Area of a triangle			5			2	2		3		4		2	2					2	
Cosine rule (for a side)		5	5								3	4						3		
Cosine rule (for an angle)						3			3											3
Sine rule						5		6								5		5		
Simultaneous equations	6	4				6		3	6		6		6		3	4				6
Boxplot	3				4		3				2		2				4			
Dotplot				4												2				
Pie chart										3										
Quartiles/median	3		3		3		3		3		3		3		3			3	3	
Semi-interquartile range			2						2		2								2	
Standard deviation		4		6		5		6		6		6		4		6	4			5
Cumulative frequency	2							2			2								1	
Probability	1		1				1		2		1		1		1		1	1		2

UNIT 3

YEAR	2000		2001		2002		2002		2003		2004		2005		2006		2007		2008	
PAPER	1	2	1	2	1	2	1	2	1	2	1	2	1	2	1	2	1	2	1	2
Simplifying fractions								2	2								1			
Add/subtract fractions	3		3			3	2		2		3				3					3
Multiply/divide fractions													2				3			
Changing the subject		2	3			2	2		3		3		3		2		3		2	
Simplifying surds				3			2		2		3				3		3			3
Rationalizing the denominator	2				2								2							
Indices	2			2	2	2			2		2	3	2		2			2	1	
Quadratic graphs	2	5	5		2		6		3		5		5		4		3		6	
Quadratic equations (factorise)							3		1							5	2			
Quadratic equations (formula)		4		4		4		4		4		4		4				4		4
Trig graphs	1		1		1				2		2		3		1		2		2	
Trig equations		3		3		3		6		3		3		3		6		3		3
Trig identities		2		2		2				2				2						2

UNIT 4: Applications of Mathematics

YEAR	2000		2001		2002		2002		2003		2004		2005		2006		2007		2008	
PAPER	1	2	1	2	1	2	1	2	1	2	1	2	1	2	1	2	1	2	1	2
Bonus						3														
Payslips				5																
Overtime	4						4		3						3					3
Commission		2						2		6			2				3			
Income tax	5							5				4			5					4
Loan repayments				6								4	4		3		6			
Credit cards													5							5
Network diagrams					2				2				2		1			1	1	
Tree diagrams					4						5									
Flowcharts		4	4					4				3				4	3		2	
Spreadsheets					5					5			2			4				4
Formulae	6		6		6		3		5		5		6		2		4		3	
Finding the mean		5		5	6			5				5						6		
Cum. freq. curve (draw)										4			4							
Cum. freq. curve (interpret)							3			1				1	4		3			
Histograms																			4	